Methods in Molecular Biology

Series Editor
John M. Walker
School of Life and Medical Sciences
University of Hertfordshire
Hatfield, Hertfordshire, AL10 9AB, UK

For further volumes:
http://www.springer.com/series/7651

Microbial Metabolic Engineering

Methods and Protocols

Edited by

Christine Nicole S. Santos and Parayil Kumaran Ajikumar

Manus Bio, Cambridge, MA, USA

Humana Press

Editors
Christine Nicole S. Santos
Manus Bio
Cambridge, MA, USA

Parayil Kumaran Ajikumar
Manus Bio
Cambridge, MA, USA

ISSN 1064-3745 ISSN 1940-6029 (electronic)
Methods in Molecular Biology
ISBN 978-1-4939-9141-9 ISBN 978-1-4939-9142-6 (eBook)
https://doi.org/10.1007/978-1-4939-9142-6

Library of Congress Control Number: 2019930843

This Humana Press imprint is published by the registered company Springer Science+Business Media, LLC, part of Springer Nature.
The registered company address is: 233 Spring Street, New York, NY 10013, U.S.A.

Preface

The field of microbial metabolic engineering has matured significantly since it was formalized almost three decades ago. The work of early academics on systematizing the engineering of microbial metabolism and biochemistry has now made its way into a booming young industry with hundreds of companies—both small and large—investing time, money, and capital into the development of microbial fermentation processes for fuels, chemicals, and complex ingredients. With this comes a renewed interest in developing new tools and methodologies for quickly and efficiently developing these microbial cellular factories.

This volume of Methods in Molecular Biology is targeted toward researchers and scientists interested in engineering and optimizing microbes (bacteria and yeast) for a variety of biotechnological applications. The content covers three main topics which roughly follow what we believe to be the biggest challenges in microbial metabolic engineering: (1) the identification of relevant biosynthetic pathways for engineering; (2) the development of genetic tools for manipulating enzymes, biosynthetic pathways, and whole genomes; and (3) the application of targeted and global systems biology tools to maximize biological understanding of the engineered systems.

Chapters 1 and 2 will take readers through two different approaches for pathway elucidation by either using metagenomics to discover enzymes and biosynthetic gene clusters in a wide array of natural hosts (Yoshikuni and Ke) or using in silico methods to build completely novel pathways by leveraging enzyme promiscuity (Tyo and coworkers).

The next several chapters cover a diverse array of tools for the manipulation of individual enzymes, whole biosynthetic pathways, and genomes in a number of microbial hosts. In Chap. 3, Yan and coworkers describe the precise design of antisense RNA elements for downregulating native gene expression. In Chap. 4, Parayil and coworkers outline a way to overexpress genes and pathways at multiple stable genetic copies using chemically inducible chemical evolution (CIChE) in bacteria. In Chap. 5, Das and coworkers give guidance on the bacterial expression and characterization of cytochrome P450s, membrane-bound enzymes which play key functions in a number of interesting natural product pathways. We revisit the theme of multicopy integration—this time in yeast—in Chap. 6 where Zhao and coworkers describe the utility of combining delta integration and CRISPR (Di-CRISPR) for the efficient incorporation of large biosynthetic pathways in *Saccharomyces cerevisiae*. The optimization of complex biosynthetic pathways can be challenging, and so we have included two chapters covering different methods for the combinatorial assembly of multigene pathways using double-stranded DNA (Chap. 7, Marillonet and Werner) or single-stranded elements (Chap. 8, De Mey and coworkers). To quickly assess the impact of genetic knockouts, Brynildsen and coworkers outline procedures for leveraging the KEIO knockout collection and phage transduction for rapid phenotyping in Chap. 9. Because the methods described thus far have been optimized primarily for traditional hosts (*Escherichia coli* and *S. cerevisiae*), we also include two chapters which outline the state-of-the-art technologies for nontraditional host organisms including the cyanobacterium, *Synechocystis* sp. PCC 6803 (Chap. 10, Peebles and coworkers), and the oleaginous yeast, *Yarrowia lipolytica* (Chap. 11, Xu and coworkers).

The final set of chapters in this book highlights the importance of gaining a deeper understanding of biology to more effectively engineer and optimize biological systems.

In many published cases, the deep characterization of the underlying cellular biochemistry afforded by metabolomics or proteomics methods has proven to be crucial for the identification and targeting of enzymatic or metabolic bottlenecks. Mass spectrometry-based methods for both metabolomics and proteomics are included in Chap. 12 (Styczynski and coworkers) and Chap. 13 (Tseng and Santos), respectively. In silico approaches can also be utilized to develop metabolic models for predicting cellular phenotype; in Chap. 14, Chandrasekaran describes a genome-scale analysis which integrates both metabolic and regulatory network models for phenotype testing. In Chap. 15, Tang and He detail the use of stable isotopic carbon substrates and flux balance analysis to further elucidate cellular metabolism, particularly for identifying central metabolic nodes, imbalances, and hidden pathways. The final chapter of the book (Chap. 16, Bachhawat and Wadhwa) deviates slightly from the theme of characterization and instead shows that there are times when a powerful high-throughput assay—in this case a visual colorimetric screen—coupled with rapid genetic diversification can also be a fruitful avenue for improving cellular phenotype.

Altogether, these 16 chapters cover a wide array of topics which should prove to be helpful for anyone faced with the task of engineering complex biological systems. We certainly hope that these chapters may serve as a guide and inspiration for the next generation of microbial metabolic engineers.

Cambridge, MA, USA

Christine Nicole S. Santos
Parayil Kumaran Ajikumar

Contents

Contributors

SANDRA J. AEDO • *Department of Chemical and Biological Engineering, Princeton University, Princeton, NJ, USA*
PARAYIL KUMARAN AJIKUMAR • *Manus Bio, Cambridge, MA, USA*
EE LUI ANG • *Metabolic Engineering Research Laboratory, Science and Engineering Institutes, Agency for Science, Technology and Research, Singapore, Singapore*
ANAND K. BACHHAWAT • *Department of Biological Sciences, Indian Institute of Science Education and Research, Mohali, S.A.S Nagar, Punjab, India*
D. BAUWENS • *Department of Biotechnology, Ghent University, Ghent, Belgium*
BRADLEY W. BIGGS • *Department of Chemical and Biological Engineering, Northwestern University, Evanston, IL, USA*
MARK P. BRYNILDSEN • *Department of Chemical and Biological Engineering, Princeton University, Princeton, NJ, USA*
SRIRAM CHANDRASEKARAN • *Department of Biomedical Engineering, University of Michigan at Ann Arbor, Ann Arbor, MI, USA*
P. COUSSEMENT • *Department of Biotechnology, Ghent University, Ghent, Belgium*
ADITI DAS • *Department of Comparative Biosciences, University of Illinois at Urbana-Champaign, Urbana, IL, USA; Department of Biochemistry, University of Illinois at Urbana-Champaign, Urbana, IL, USA; Beckman Institute for Advanced Science and Technology, University of Illinois at Urbana-Champaign, Urbana, IL, USA; Division of Nutritional Science, Neuroscience Program, Department of Bioengineering, University of Illinois at Urbana-Champaign, Urbana, IL, USA*
MARJAN DE MEY • *Department of Biotechnology, Ghent University, Ghent, Belgium*
JAKE ENGEL • *Department of Chemical, Biochemical and Environmental Engineering, University of Maryland Baltimore County, Baltimore, MD, USA*
LIAN HE • *Department of Energy, Environmental and Chemical Engineering, Washington University, St. Louis, MO, USA*
CHRIS HENRY • *Mathematics and Computer Science Division, Argonne National Laboratory, Argonne, IL, USA*
BENJAMIN HOLDRIDGE • *Department of Chemical, Biochemical and Environmental Engineering, University of Maryland Baltimore County, Baltimore, MD, USA*
WHITNEY HOLLINSHEAD • *Department of Energy, Environmental and Chemical Engineering, Washington University, St. Louis, MO, USA*
HANNAH HUFF • *Department of Chemistry, University of Illinois at Urbana-Champaign, Champaign, IL, USA*
JAMES JEFFRYES • *Department of Chemical and Biological Engineering, Northwestern University, Evanston, IL, USA; Mathematics and Computer Science Division, Argonne National Laboratory, Argonne, IL, USA*
JING KE • *U.S. Department of Energy Joint Genome Institute, Lawrence Berkeley National Laboratory, Walnut Creek, CA, USA*
YOUYUN LIANG • *Metabolic Engineering Research Laboratory, Science and Engineering Institutes, Agency for Science, Technology and Research, Singapore, Singapore*
AARON M. LOVE • *Manus Bio, Cambridge, MA, USA*

HELENA R. MA • *Department of Chemical and Biological Engineering, Princeton University, Princeton, NJ, USA*
J. MAERTENS • *Department of Biotechnology, Ghent University, Ghent, Belgium*
SYLVESTRE MARILLONNET • *Department of Cell and Metabolic Biology, Leibniz-Institut für Pflanzenbiochemie, Halle, Germany*
DEMETRIOS MAROUTSOS • *Department of Biochemistry, University of Illinois at Urbana Champaign, Urbana, IL, USA*
APRIL M. MIGUEZ • *Georgia Tech School of Chemical & Biomolecular Engineering, Atlanta, GA, USA*
CHRISTIE ANN MARIE PEEBLES • *Department of Chemical and Biological Engineering, Colorado State University, Fort Collins, CO, USA; Cell and Molecular Biology Graduate Program, Colorado State University, Fort Collins, CO, USA*
CHRISTINE NICOLE S. SANTOS • *Manus Bio, Cambridge, MA, USA*
JACOB SEBESTA • *Department of Chemical and Biological Engineering, Colorado State University, Fort Collins, CO, USA*
SHUOBO SHI • *Metabolic Engineering Research Laboratory, Science and Engineering Institutes, Agency for Science, Technology and Research, Singapore, Singapore*
MCKENZIE L. SMITH • *Georgia Tech School of Chemical & Biomolecular Engineering, Atlanta, GA, USA*
JONATHAN STRUTZ • *Department of Chemical and Biological Engineering, Northwestern University, Evanston, IL, USA*
MARK P. STYCZYNSKI • *Georgia Tech School of Chemical & Biomolecular Engineering, Atlanta, GA, USA*
YINJIE J. TANG • *Department of Energy, Environmental and Chemical Engineering, Washington University, St. Louis, MO, USA*
HSIEN-CHUNG TSENG • *Manus Bio, Cambridge, MA, USA*
KEITH E. J. TYO • *Department of Chemical and Biological Engineering, Northwestern University, Evanston, IL, USA*
MANISHA WADHWA • *Department of Biological Sciences, Indian Institute of Science Education and Research, Mohali, S.A.S Nagar, Punjab, India*
JIAN WANG • *College of Engineering, University of Georgia, Athens, GA, USA*
ALLISON WERNER • *Cell and Molecular Biology Graduate Program, Colorado State University, Fort Collins, CO, USA*
STEFAN WERNER • *Nambawan Biotech GmbH, Halle, Germany*
LYNN WONG • *Department of Chemical, Biochemical and Environmental Engineering, University of Maryland Baltimore County, Baltimore, MD, USA*
PENG XU • *Department of Chemical, Biochemical and Environmental Engineering, University of Maryland Baltimore County, Baltimore, MD, USA*
YAJUN YAN • *School of Chemical, Materials and Biomedical Engineering, College of Engineering, University of Georgia, Athens, GA, USA*
YAPING YANG • *College of Engineering, University of Georgia, Athens, GA, USA*
YASUO YOSHIKUNI • *U.S. Department of Energy Joint Genome Institute, Lawrence Berkeley National Laboratory, Walnut Creek, CA, USA*
RUIHUA ZHANG • *College of Engineering, University of Georgia, Athens, GA, USA*
HUIMIN ZHAO • *Metabolic Engineering Research Laboratory, Science and Engineering Institutes, Agency for Science, Technology and Research, Singapore, Singapore; Department of Chemical and Biomolecular Engineering, University of Illinois at Urbana-Champaign, Urbana, IL, USA*

Chapter 1

Pathway and Gene Discovery from Natural Hosts and Organisms

Jing Ke and Yasuo Yoshikuni

Abstract

Information in public sequence databases on the genomes and metagenomes of microbes and plants has grown rapidly. In conjunction with technological developments in computational identification of biosynthetic gene clusters, molecular biology, synthetic biology, and analytical tools, this has revealed genes for enzymes with optimal and targeted function, as well as a rich pool of uncharacterized metabolic pathways. This chapter discusses different approaches to discovery of genes and metabolic pathways in microbes and plants in nature, such as genomic mining, transcriptome analysis, and metabolite profiling.

Key words Pathway, Gene cluster, Enzyme, Gene sequence, Metagenomics, Gene library, Computational identification, Metabolome, Transcriptome

1 Introduction

For much of its history, the search to discover new genes and metabolic pathways from microbes and plants was driven at the metabolite level, either by bioactivity-guided fractionation from fermentation broth, or isolation of chromatographically resolvable metabolites with promising spectroscopic properties. With the rapid growth of genome and metagenome databases, in conjunction with the development of bioinformatics, analytical tools, and synthetic biology, the search for new genes and pathways is becoming more amenable to exploration at the genome level. Although discovery of genes and pathways still requires large-scale genomic data mining and experimental verification, a number of bioinformatics tools (*see* Subheading 2) and experimental protocols are now available to pose sophisticated queries to data repositories. From the genome sequence data currently available in public databases, it is already clear that there is potential for discovery of novel genes and pathways for secondary metabolites in many microbes and plants. The easy and inexpensive access to genomic sequence data

Christine Nicole S. Santos and Parayil Kumaran Ajikumar (eds.), *Microbial Metabolic Engineering: Methods and Protocols*, Methods in Molecular Biology, vol. 1927, https://doi.org/10.1007/978-1-4939-9142-6_1,

resulting from next-generation sequencing technologies [1] has created a new paradigm for genome mining. Biosynthetic gene clusters (BGCs) that are identified computationally can be cloned and expressed in heterologous systems, which can then be monitored for production of novel metabolites [2]. At the same time, single-cell genome sequencing and metagenomics are opening up access to uncultivable, uncharted branches on the tree of life. Furthermore, synthetic biology allows BGCs to be redesigned for heterologous expression in pre-engineered hosts, which will ultimately enable construction of standardized high-throughput platforms for gene and pathway discovery. It is now possible to carry out experimental characterization of hundreds of such secondary metabolites, using the high-throughput techniques of mass spectrometry.

2 Materials

Selected tools for mining secondary metabolite biosynthesis gene clusters are:

1. 2metDB (http://secmetdb.sourceforge.net/).
2. antiSMASH. (http://antismash.secondarymetabolites.org).
3. BAGEL (http://bagel.molgenrug.nl/).
4. CASSIS and SMIPS (https://sbi.hki-jena.de/cassis/).
5. CLUSEAN (https://bitbucket.org/tilmweber/clusean).
6. ClusterFinder (https://github.com/petercim/ClusterFinder).
7. ClustScan Professional (http://bioserv.pbf.hr/cms/index.php?page=clustscan).
8. eSNaPD//environmental Surveyor of Natural Product Diversity (http://esnapd2.rockefeller.edu/).
9. EvoMining (http://148.247.230.39/newevomining/new/evomining_web/index.html).
10. FunGeneClusterS (https://fungiminions.shinyapps.io/FunGeneClusterS).
11. MIDDAS-M (http://133.242.13.217/MIDDAS-M).
12. MIPS-CG (http://www.fung-metb.net/).
13. NaPDoS//Natural Products Domain Seeker (http://napdos.ucsd.edu/).
14. PRISM (http://magarveylab.ca/prism)/GNP (http://magarveylab.ca/gnp/#!/genome).
15. SMURF/Secondary Metabolite Unknown Region Finder (http://jcvi.org/smurf/index.php).
16. NCBI BLAST (https://blast.ncbi.nlm.nih.gov/Blast.cgi).

3 Methods

3.1 Discovery of Enzymes with Optimal Activity

Protein engineering approaches including rational design, directed evolution, and computational design [3] have been routinely applied to generate enzyme variants with more optimal function. However, these approaches are often cumbersome and almost always require high-resolution structural information and/or high-throughput functional assays. Recent expansion of genomic information available through public databases and rapid decrease in costs of DNA synthesis are improving researchers' ability to mine for enzymes with desired functions [4, 5]. This mining approach allows us to evaluate significantly larger sequence spaces using a relatively small sample sizes.

1. To obtain genomic enzyme ortholog (GEO) sequences, the native sequence is input as a query for a homologous sequence search using HMMER3's online server (http://www.ebi.ac.uk/Tools/hmmer/) [6].
2. The resulting GEO sequences are aligned using a multiple sequence comparison by log-Expectation (MUSCLE) multiple sequence alignment (MSA) program [7] (*see* **Note 1**).
3. A phylogenetic tree of these GEOs is generated using methods such as Neighbor-Joining and UPGMA with a Jukes-Cantor genetic distance model. The resulting tree is visualized using the iTOL online tool (http://itol.embl.de/) in the circular tree mode and rooted at native sequence [8, 9].
4. To select GEOs for synthesis, the MSA is filtered using the CD-HIT online server (http://weizhongli-lab.org/cdhit_suite/cgi-bin/index.cgi) (*see* **Note 2**) with an identity cutoff [10, 11]. The percentage of identify cutoff is determined based on the tractable number of sampling sizes for synthesis and the throughput of subsequent biochemical characterization.
5. (Optional) If structural information for a member of the sequence family is available, a homology model of each sequence is generated using Rosetta Comparative Modeling [12] (*see* **Note 3**). At this point, the reaction intermediate is placed into the active site. For each model, the GEO with the lowest energy for overall protein and protein–ligand interface is selected.
6. Genes encoding each selected GEO are subjected to a BOOST online server (*see* **Note 4**). The BOOST web tool (https://boost.jgi.doe.gov) [13] allows (a) codon optimization of each GEO for expression in heterologous hosts such as *E. coli*, (b) elimination of sequences (e.g., repeats, high/low local/global GC content, and secondary structures) that are difficult to synthesize, and (c) partition if the final constructs are too

large to synthesize in a single step. Resulting sequences are ordered through commercial vendors.

7. When the GEO with the most optimal function is identified through biochemical characterization, this process can be repeated from **step 4**, searching through the clade to which the GEO belongs to identify a GEO variant with even more optimal function.

3.2 Discovery of Enzymes with Targeted Novel Biochemical Activity

Activity-based studies remain crucial for identifying enzymes with novel activity. Selecting appropriate organisms and environmental communities enriched in the targeted activities and designing screens to identify enzymes with desired characteristics are key in the field [14]. Functional genomics approaches including sequence-based function annotation and differential expression analyses are often useful to downselect the enzymes to be tested for a desired function.

1. Develop and employ selective functional screening to identify the specific phenotypes associated with the novel activities.
2. Identify conditions under which specific phenotypes are significantly more induced or suppressed compared with a standard condition.
3. Isolate RNA samples from organisms subjected to these conditions, and create libraries (*see* **Note 5**).
4. Analyze the differential expression of mRNAs for these conditions using the Trinity [15] (*see* **Note 6**) and/or Cufflinks [16] program packages.
5. Calculate FPKM (fragments per kilobase of exon per million fragments mapped) for all gene expression (*see* **Note 7**), and rank the semitargeted genes based on their correlations to the phenotypic changes.
6. Subject the tractable number of genes that are highly ranked through RNA-seq analyses at **steps 4** and **5** to the BOOST online server, have them synthesized by commercial vendors, and clone them into an appropriate vector system.
7. To confirm novel enzyme activity in the host, use an appropriate functional assay to confirm acquisition of desired novel biochemical activity in the host expressing the genes synthesized in **step 6**.

3.3 Discovery of Novel Pathways in Bacteria and Fungi (with Gene Cluster)

In bacteria and fungi, functionally relevant genes composing metabolic pathways are often clustered near each other within genomes. Therefore, discovery of gene clusters with novel functions can be facilitated through genome-wide bioinformatics analyses [17]. Here we describe the general method for identifying those gene clusters.

1. Screen the microbes with the desired phenotype using appropriate biological assays.
2. (Optional) If the genome sequence for the microbe selected in **step 1** is not available, isolate the genomic DNA to obtain a draft genome sequence. For bacteria, the sequence assembly and annotation can be carried out through the JGI's genome portal (https://img.jgi.doe.gov). For fungi, RNA also needs to be isolated and sequenced to build accurate gene models.
3. To identify secondary metabolite biosynthetic gene clusters, there are several bioinformatics tools such as AntiSMASH (Antibiotics and Secondary Metabolite Analysis Shell) [18] (*see* **Note 8**) and ClusterFinder prediction algorithm (*see* **Note 9**). To identify pathways other than secondary metabolite biosynthetic pathways, one useful tool is Basic Local Alignment Search Tool (BLAST) (https://blast.ncbi.nlm.nih.gov/Blast.cgi) [19]. Doing a query on BLAST, followed by a conserved neighborhood search through the JGI's genome portal, will provide a protein sequence that is likely relevant to the function of interest. These evolutionarily related hook proteins can be identified through the differential expression analyses discussed in Subheading 3.2. Comparing the configuration of gene clusters from multiple species could provide critical insights into pathways of interest. If biochemical steps are partially understood, the known information can be used to evaluate the completeness of the gene clusters as well.
4. Clone the gene clusters selected in **step 3** into a vector. The vector system based on bacterial artificial chromosome (BAC) is often favorable for cloning large gene clusters. The BAC vector can hold up to 200 kbp. Artificial promoters can be added to the constructs to help drive transcription. A synthetic biology approach (refactoring and synthesis) also helps with expressing functional gene clusters in heterologous hosts.
5. Function of the gene clusters can be determined through biochemical screenings.

3.4 Discovery of Pathways in Plants

Unlike genes in bacteria and fungi, genes involved in plant metabolic and regulatory pathways are not physically clustered. Differential coexpression analysis of transcriptomes, when combined with metabolite profiling, is a promising approach to discovering novel pathways in plants [20, 21].

1. To preliminarily explore a promising pathway in a particular plant species, perform tissue and media extraction for transcriptome and metabolome analyses (*see* **Note 10**), normally by RNA-seq and HPLC-MS/MS, respectively.
2. For metabolome analysis, collect mass spectrometry data in both positive and negative ion modes and analyze using

MassHunter Qualitative Analysis software (Agilent) and XCMS [22, 23].

3. For untargeted metabolomics, convert Agilent MassHunter data files to mzXML format using trapper (Seattle Proteome Center, http://tools.proteomecenter.org/software.php). Preprocess and analyze grouped mzXML files by XCMS using a standard XCMS workflow (http://metlin.scripps.edu/xcms/xcmsPreprocess.pdf). A sample R script for data preprocessing and report generation is given as:

```
library(xcms)
xset<-xcmsSet()
xset<-group(xset)
xset2<-retcor(xset,family="s",plottype="m")
xset2<-group(xset2)
xset3<-retcor(xset2,family="s",plottype="m")
xset3<-group(xset3)
xset4<-retcor(xset3,family="s",plottype="m")
xset4<-group(xset4,bw=10)
xset5<-fillPeaks(xset4)
reporttab<-
diffreport(xset5,"WT_+Psta","Mut_+Psta","WT_v_Mut_
+Psta",500)
```

The resulting report, containing a mass peak list with m/z values, peak intensity fold change, statistical significance (*see* **Note 11**), retention times, and extracted peak intensities, is used to identify compounds of interest. The initial peak list is filtered using various cutoff criteria (*see* **Note 12**) to yield a narrower list for further analysis.

4. (Optional) Isolate microsomal protein from plant tissue and detect desired enzyme activity in this fraction.

5. Isolate total RNA from plant tissue. For the RNA-seq data analysis pipeline, four steps are required: (1) perform quality control operations using Trimmomatic [24]; (2) align reads to the reference genome using TopHat [15]; (3) assemble transcripts using the Cufflinks [15] package, enabling both multiread correction and fragment bias correction, and quantify them using units of fragments per kilobase of exon per million mapped fragments (FPKM); (4) extract the expression levels of the genes, as annotated in Phytozome [25], to a spreadsheet using a custom Python script.

6. Prepare a cDNA library from extracted mRNA using the SuperScript III First-Strand Synthesis System (Invitrogen) according to the manufacturer's instructions. If accurate gene models are available, cDNA can be directly synthesized with optimal codon usage for expression in heterologous host systems.

7. Transform cDNAs from **step 2** into model host species such as *E. coli*, *S. cerevisiae*, and/or *Nicotiana tabacum* for functional characterization. The expression system using *Nicotiana tabacum* is particularly useful, as several genes are transformed and expressed simultaneously.

4 Notes

1. The MUSCLE algorithm is delivered as a command-line program called muscle. If run under Linux or Unix, the program works at a shell prompt. If run under Windows, it should be in a command window. A range of options allows the user to optimize accuracy or speed, or to choose some compromise between the two. Default parameters are those that gave the best average benchmark accuracy in their current software tests.
2. Alternatively, the program can be downloaded for offline use.
3. The alignment file must be passed as input, and the format must be specified as grishin, modeller, vie, hhsearch, clustalw, or fasta.
4. BOOST is available at no cost to noncommercial enterprises, and commercial use is available through the Innovations and Partnerships Office of Lawrence Berkeley National Laboratory (ipo@lbl.gov). Every tool in BOOST supports the input and output of sequences in FASTA, GenBank, synthetic biology open language (SBOL), and simple comma-separated values (CSV). BOOST prefers GenBank and SBOL sequence exchange formats, which support sequence feature annotation. The formats of CSV and FASTA do not support feature annotations. Also, if these formats are used, the user must enable the "Auto-Annotate" feature that annotates each sequence as a coding sequence ("CDS") if the sequence's length is a multiple of 3 bp in order to take advantage of BOOST's codon-juggling and sequence-polishing features.
5. While there are various RNA-Seq applications and protocols, most follow the basic strategy of isolating RNA (such as with poly-dT to pull down mRNA), converting it to DNA, and then adding adaptor sequences to generate a library.
6. The Trinity program is useful when the reference genome sequences are not available.
7. FPKM is not a method, but simply a unit of the RNA-Seq expression unit. The formula is FPKM = (CDS fragment count × 109)/(CDS length × total mapped read count).
8. The input for antiSMASH should be an annotated nucleotide file in the format of GenBank, EMBL, or FASTA. The analysis is selected by default, and a phylogenetic tree is constructed of

each gene together with the (max. 100) sequences of the secondary metabolism gene families seed alignment.

9. The ClusterFinder algorithm detects gene clusters of both known and unknown classes. The source code is available from the GitHub repository (https://github.com/petercim/ClusterFinder).
10. Metabolome analysis covers identification and quantification of intracellular and extracellular metabolites with molecular mass lower than 1000 Da, using different analytical techniques. It also varies depending on whether it is used to characterize predefined metabolites (target analysis) or scan all detectable metabolites (metabolite profiling).
11. Evaluated by p value, two-tailed unequal variance Student's t-test.
12. For example, p less than 0.05, fold change greater than 2, average peak intensity for either experimental condition greater than 2×10^5.

References

1. Didelot X, Bowden R, Wilson DJ, Peto TE, Crook DW (2012) Transforming clinical microbiology with bacterial genome sequencing. Nat Rev Genet 13:601–612
2. Rutledge PJ, Challis GL (2015) Discovery of microbial natural products by activation of silent biosynthetic gene clusters. Nat Rev Microbiol 13:509–523
3. Chica RA, Doucet N, Pelletier JN (2005) Semi-rational approaches to engineering enzyme activity: combining the benefits of directed evolution and rational design. Curr Opin Biotechnol 16:378–384
4. Mak WS, Tran S, Marcheschi R, Bertolani S, Thompson J, Baker D et al (2015) Integrative genomic mining for enzyme function to enable engineering of a non-natural biosynthetic pathway. Nat Commun 6. https://doi.org/10.1038/ncomms10005
5. Heins RA, Cheng X, Nath S (2014) Phylogenomically guided identification of industrially relevant GH1 beta-glucosidases through DNA synthesis and nanostructure-initiator mass spectrometry. ACS Chem Biol 9:2082–2091
6. Finn RD, Clements J, Eddy SR (2011) HMMER web server: interactive sequence similarity searching. Nucleic Acids Res 39: W29–W37
7. Edgar RC (2004) MUSCLE: multiple sequence alignment with high accuracy and high throughput. Nucleic Acids Res 32:1792–1797
8. Letunic I, Bork P (2007) Interactive tree of life (iTOL): an online tool for phylogenetic tree display and annotation. Bioinformatics 23:127–128
9. Letunic I, Bork P (2011) Interactive tree of life v2: online annotation and display of phylogenetic trees made easy. Nucleic Acids Res 39:475–478
10. Huang Y, Niu B, Gao Y, Fu L, Li W (2010) CD-HIT suite: a web server for clustering and comparing biological sequences. Bioinformatics 26:680–682
11. Kearse M, Moir R, Wilson A, Stones-Havas S, Cheung M, Sturrock S et al (2012) Geneious basic: an integrated and extendable desktop software platform for the organization and analysis of sequence data. Bioinformatics 28:1647–1649
12. Song Y, DiMaio F, Wang RY, Kim D, Miles C, Brunette TJ et al (2013) High-resolution comparative modeling with RosettaCM. Structure 21:1735–1742
13. Oberortner E, Cheng J, Hillson NJ, Deutsch S (2016) Streamlining the design-to-build transition with build-optimization software tools. ACS Synth Biol. https://doi.org/10.1021/acssynbio.6b00200
14. Enquist-Newman M, Faust AME, Bravo DD, Santos CN, Raisner RM, Hanel A (2014)

Efficient ethanol production from brown macroalgae sugars by a synthetic yeast platform. Nature 505:239–243

15. Trapnell C, Roberts A, Goff L, Pertea G, Kim D, Kelley DR et al (2012) Differential gene and transcript expression analysis of RNA-seq experiments with TopHat and Cufflinks. Nat Protoc 7:562–578
16. Trapnell C, Hendrickson DG, Sauvageau M, Goff L, Rinn JL, Pachter L (2013) Differential analysis of gene regulation at transcript resolution with RNA-seq. Nature Biotech 31:46–53
17. Wargacki AJ, Leonard E, Win MN, Regitsky DD, Santos CNS, Kim PB et al (2012) An engineered microbial platform for direct biofuel production from brown macroalgae. Science 335:308–313
18. Weber T, Blin K, Duddela S, Krug D, Kim HU, Bruccoleri R et al (2015) AntiSMASH 3.0—a comprehensive resource for the genome mining of biosynthetic gene clusters. Nucleic Acids Res 43:W237–W243
19. Marchler-Bauer A, Lu S, Anderson JB, Chitsaz F, Derbyshire MK, DeWeese-Scott C et al (2011) CDD: a conserved domain database for the functional annotation of proteins. Nucleic Acids Res 39(suppl 1):D225–D229
20. Malcolmson J, Meek SJ, Sattely ES, Schrock RR, Hoveyda AH (2008) Highly efficient molybdenum-based catalysts for enantioselective alkene metathesis. Nature 456:933–937
21. Klein P, Sattely ES (2015) Two cytochromes P450 catalyze S-heterocyclizations in cabbage phytoalexin biosynthesis. Nat Chem Biol 11:837–839
22. Smith CA, Want EJ, O'Maille G, Abagyan R, Siuzdak G (2006) XCMS: processing mass spectrometry data for metabolite profiling using nonlinear peak alignment, matching, and identification. Anal Chem 78:779–787
23. De Vos RC, Moco S, Lommen A, Keurentjes JJ, Bino RJ, Hall RD (2007) Untargeted large-scale plant metabolomics using liquid chromatography coupled to mass spectrometry. Nat Protoc 2:778–791
24. Bolger AM, Lohse M, Usadel B (2014) Trimmomatic: a flexible trimmer for Illumina sequence data. Bioinformatics 30:2114–2120
25. Goodstein DM, Shu S, Howson R, Neupane R, Hayes RD, Fazo J et al (2012) Phytozome: a comparative platform for green plant genomics. Nucleic Acids Res 40:D1178–D1186

Chapter 2

Metabolic In Silico Network Expansions to Predict and Exploit Enzyme Promiscuity

James Jeffryes, Jonathan Strutz, Chris Henry, and Keith E. J. Tyo

Abstract

There is a growing consensus that enzymes are capable of catalyzing not just one canonical reaction but entire families of related reactions. These capacities often go unnoticed in the enzyme's native context but can become apparent in engineered metabolism when the enzyme is exposed to novel substrates or high concentrations of pathway intermediates. This chapter describes how to use metabolic in silico network expansion (MINE) databases to predict novel biotransformations and their resulting metabolites. In particular, searching MINEs by structural similarity or with metabolomics data allows scientists to detect, exploit, or avoid these predicted transformations.

Key words Enzyme promiscuity, Untargeted metabolomics, Metabolite damage, Feature annotation

1 Introduction

It now appears that substrate promiscuity is quite common in metabolism, and this promiscuity can form the basis for engineering reactions on nonnative substrates. Unfortunately, enzyme promiscuity has historically been underreported, largely due to the fact that it is very difficult to study [1, 2]. Studying substrate-level enzyme promiscuity has been experimentally challenging because (a) it is difficult to study specific substrate–enzyme interactions by chemical analytics in the complicated cellular milieu, (b) promiscuity is not amenable to study by genetic methods, and (c) the necessary substrates needed for in vitro enzyme assays can be difficult to obtain.

Like substrate promiscuity, metabolite damage (the enzymatic conversion of a metabolite to a nonuseful compound) has only recently been recognized as a significant feature of metabolic networks [3, 4]. Ribulose bisphosphate carboxylase's (Rubisco) promiscuous reaction with O_2, rather than CO_2, is a prominent example [5]. Likewise, L-malate dehydrogenase can reduce

Christine Nicole S. Santos and Parayil Kumaran Ajikumar (eds.), *Microbial Metabolic Engineering: Methods and Protocols*, Methods in Molecular Biology, vol. 1927, https://doi.org/10.1007/978-1-4939-9142-6_2,

α-ketoglutarate to L-2-hydroxyglutarate, and in humans, it produces grams per day of the damaged metabolite that must be "repaired" by other enzymes [6]. These cases are widespread, as a recent report estimated up to 37% of *E. coli* enzymes have been categorized as "generalists," that is, being documented to catalyze more than one reaction [7]. Heterologous enzymes are therefore very likely to catalyze damaging reactions.

Damage-prone metabolites tend to be chemically similar to the damaging enzyme's canonical substrate. Because the metabolites are chemically similar, they can fit in the active site pocket, allowing the damaging reaction to occur. Additionally, many metabolites are susceptible to spontaneous reactions [8]. These damaging/promiscuous reactions in themselves can result in a loss of valuable resources, because the carbon is converted into a form the cell cannot immediately use. Furthermore, these damaged metabolites can act as enzyme inhibitors in the metabolic network [9].

To address this unmet need, we have developed metabolic in silico network expansions (MINEs) [10] as an attempt to anticipate the promiscuous reactions that are possible, either for biosynthetic goals or to identify toxic mechanisms. MINEs take known metabolites (from a database that can be organism specific or not) and predict the many ways enzymatic and spontaneous chemistries could modify those compounds. To predict reactions, we use the Biochemical Network Integrated Computational Explorer (BNICE). BNICE represents molecules by the bonds between the atoms in a bond electron matrix (BEM) and represents reactions by adding and subtracting bonds from the BEM. For a complete description of the BNICE algorithm, please *see* previous studies [11–13].

MINEs solve specific needs for metabolic engineering and more broadly for understanding metabolism. The compounds in the MINE can be searched by chemical structure and a range of other compound attributes. The MINE contains both experimentally observed and computationally predicted compounds (Subheading 3.1). MINEs can take a known metabolic compound and predict many possible products that could be formed from promiscuous reactions (Subheading 3.2). Secondly, if one discovers a molecule not previously associated with metabolism, MINEs can be used to propose what known metabolite might be the molecule's precursors (Subheading 3.3). Finally, MINEs can assist the annotation of unknown metabolites in high-resolution mass spectrometry data by suggesting likely structures (Subheading 3.4). In the same step, MINEs can propose the precursor, as in Subheading 3.3. MINEs are an open-source database with a user-friendly web interface and an API to facilitate larger efforts.

2 Materials

The MINEs website can be found at minedatabase.mcs.anl.gov. The API can be found at github.com/JamesJeffryes/MINE-API. To input molecular structures, there is a chemical drawing option on the MINE website. Otherwise, molecules can be encoded in mol or sdf files. Mol and sdf files can be obtained from many metabolic databases or exported from common chemistry software, like ChemDraw®. Use of the metabolomics tools requires a list of mass–charge ratios of interest. This information can be obtained from several popular metabolomics workflows such as XCMS [14], MZMINE2 [15], or MS-DIAL [16].

3 Methods

3.1 Basic Overview of MINE Searching Interface

1. Go to the MINEs website.
2. Select the "Structure Search" tab at the top of the page (*see* **Note 1**).
3. On the following page (Fig. 1), there will be displayed a white space for drawing your molecule. The left-hand column can be used to add bonds (single, double, etc.). Click on the bond wanted, and click in the white space to add a bond.
4. The right-hand column contains common atoms found in biological systems. As the molecule is being built, it will assume all atoms are C or H. To change to other atoms, click on the atom type of interest, and then click on the C or H that needs converted to the desired atom type.
5. After the molecule has been entered, there are several search parameters that should be adjusted appropriately. First, the Database (in the upper left of the web page) can be changed. Four databases are currently available. If you are interested in the metabolism or *E. coli* or yeast, there are organism specific databases for each, EcoCyc [17] for *E. coli* and Yeast Metabolomc Database (YMDB) for yeast [18]. If you are working with another organism or do not know the organism, the Kyoto Encyclopedia of Genes and Genomes (KEGG) [19] has a large, diverse set of molecules from many species.
6. The Search Type is another parameter (upper right of web page). There are three options: (1) "Exact," which searches only for the exact matching molecule (2) "Similarity," which compares the *overall* similarity between two different molecules, and (3) "Substructure," which returns molecules that contain the largest common substructure between the two different molecules (*see* **Note 2**).

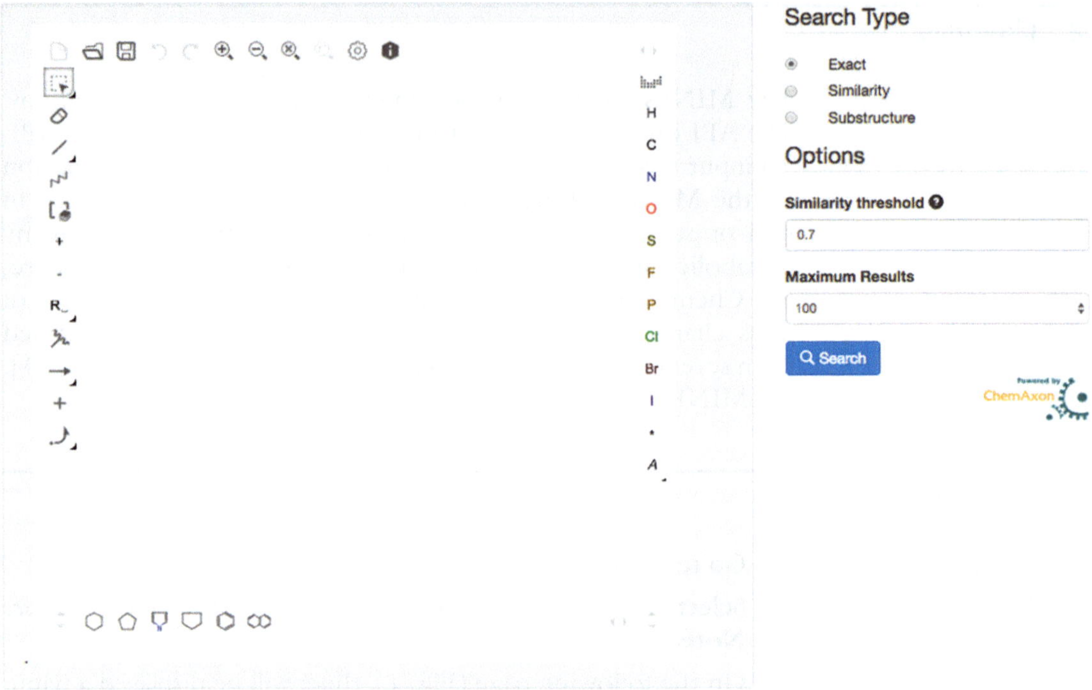

Fig. 1 Compound structure search page. This is the drawing tool that can be used to input molecular structures (e.g., Subheading 3.1, **step 3**)

7. If "Similarity" search is chosen, a minimum similarity can be imposed based on the Tanimoto coefficient [20]. A value of 1 is an exact match, and a value of 0 means no similarity is required.
8. Finally, a maximum number of molecules to be returned can be chosen (default is 100).
9. After clicking "Search," a results web page is displayed (Fig. 2). Molecules are listed in decreasing similarity or substructure commonality. The "MINE Id" is a standardized number used to reference database compounds. "Img" shows images of the chemical structures of the similar compounds, "Name" lists the names (when available), and "Formula" shows the molecular formulas on the far right.
10. If the user clicks on "MINE Id," a web page will open for the specific molecule (Fig. 3). This web page has three tabs: an "Overview," "Products," and "Reactants."
11. The "Overview" page shows basic information about the compound, including the structure, formula, exact mass, electrostatic charge at pH 7, and two different strings that can be used to define the molecular structure: InChIKey [21] and SMILES [22]. It also shows the natural product likeness (NP likeness) score [23], an indication of the degree of structural similarity to known natural products which ranges between −3 and

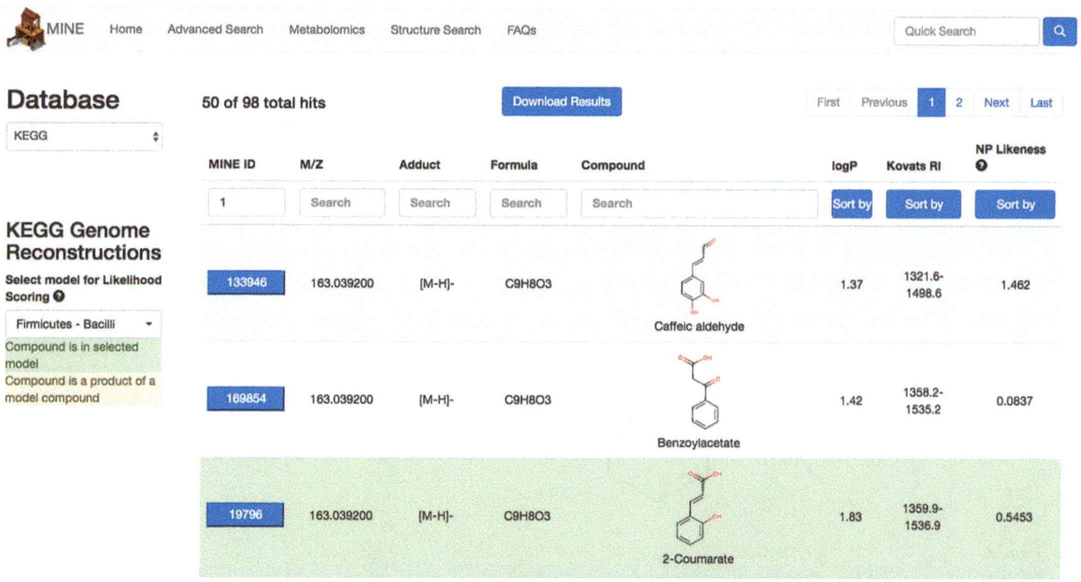

Fig. 2 MINEs search results page. This is an example of the page that is returned after a search (e.g., Subheading 3.1, **step 9**. Subheading 3.2, **step 2**, Subheading 3.3, **step 2**, and Subheading 3.4, **step 11**)

3. The octanol–water partition coefficient (logP), and the predicted Kovats retention index for gas chromatographic separation [24] are both displayed as well. Below this, a list of the enzymes (using Enzyme Commission numbers) that are known to act on the substrate or produce it as a product (if known) and the alternate names for the compound (if there are alternates) are shown. The "Databases" section cross-references the compound with several other compound databases KEGG [19], NIKKAJI [now J-Global], Chemical Abstract Services [CAS]/Sigma-Aldrich, Chemical European Bioinformatics Institute (ChEBI) [25], Model SEED [26], Protein Databank Chemical Compounds Database (PDB-CCD) [27], PubChem [28], LipidBank [29], KNApSAcK [30], Human Metabolome Database (HMDB) [31], and LIPIDMAPS [32]). The last information field contains the name of the metabolic pathway in which the compound occurs (if known).

12. The "Products" and "Reactants" tabs at the top of the web page enumerate all the MINE reactions that contain the compound as a product or reactant. In each row, it first lists the generalized reaction rule (or operator) that is consistent with the known reaction or predicts the hypothetical reaction. The rest of the column shows the reaction equation. The user can click on the reaction rule and reactants/products in each row for additional information.

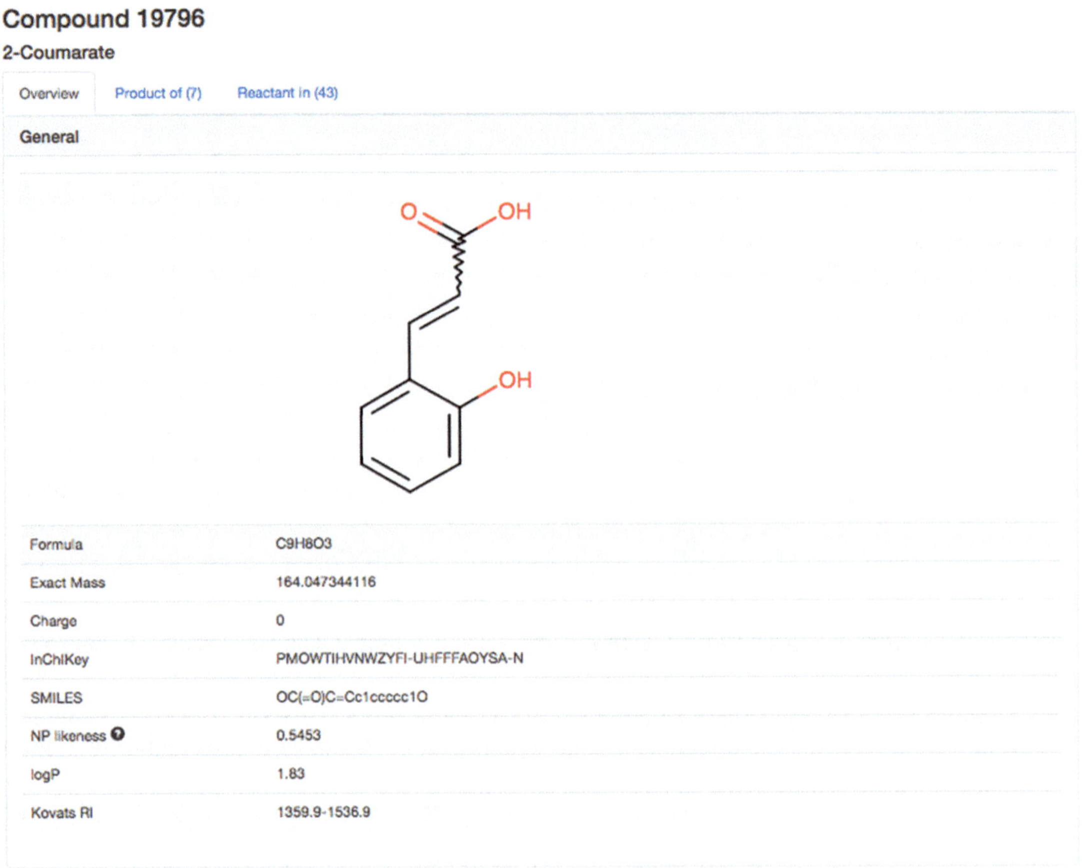

Fig. 3 Molecule characteristics page. This is an example of the page that is returned for an individual compound (either experimentally verified or predicted compound) in Subheading 3.1, **step 10**, Subheading 3.2, **step 2**, Subheading 3.3, **step 2**, and Subheading 3.4, **step 14**

13. On the results page (**step 9**), results can be downloaded as a comma-separated values (csv) file. These files contain each compound returned by the search with columns as follows: SMILES string, NP likeness score, "Generations" where "0" is a known metabolite, and "1" is a product of one promiscuous reaction of a known metabolite (*see* **Note 3**), MINE Id, InChI key, accurate mass of the molecular formula, molecular formula, and the "_id" which is a hashed internal identifier.

3.2 Searching for Possible Derivatives of a Compound of Interest

Once the user has done structure searches and interpreted the returned results, identifying possible reactions that might consume known metabolites is straightforward.

1. Search for the known metabolite that you hypothesize might be consumed in an unknown reaction using **steps 1–9** of Subheading 3.1. (Using an "Exact" search in **step 6** will simplify the search.)

2. On the results screen (Subheading 3.1, **step 9**) the metabolite entered should be shown. Clicking on the "MINE Id" will bring the user to the compound's page. Clicking on "Reactant in . . ." will display all MINE reactions that use this compound as a reactant. Validated and proposed reactions can be viewed as in Subheading 3.1, **step 12**.

3.3 Searching for Reactions that Produce a Compound of Interest

For metabolic engineering purposes, one may desire to find a reaction that produces a compound that has not been annotated in biology. For metabolomics scientists, identifying the reaction that produced a preliminarily annotated hypothetical compound may be useful to add confidence to the prediction. A procedure as in Subheading 3.2 can be used to carry out this search as well.

1. Search for the known metabolite that you hypothesize might be consumed in an unknown reaction using **steps 1–9** of Subheading 3.1. In this case, it may be advantageous to use "Similarity" because the specific compound may not be in the database, and chemical analogues that may be informative can be easily found with the "Similarity" search.
2. On the results screen (Subheading 3.1, **step 9**), the metabolite (or similar metabolites if using "Similarity") entered should be shown. Clicking on the "MINE Id" will bring the user to the compound's page. Clicking on "Products in . . ." will display all MINE reactions that use this compound as a reactant. Validated and proposed reactions can be viewed as in Subheading 3.1, **step 12**.

3.4 Search for Candidate Structures by MS Accurate Mass

1. Go to the MINEs website.
2. Select the "Metabolomics" tab at the top of the page.
3. On the following page (titled "MS Adduct Search") will be used to search the MINE compounds by mass.
4. Individual *m/z* ratios for intact ions can be entered into the "Peak M/Z" field. Multiple *m/z* can be entered on separate lines in the field. Alternately, a text file can be uploaded for bulk searching.
5. The Electrospray Ionization mode (positive or negative) must be selected in the Charge field.
6. The "Adducts" list provide common adducts that form during the ionization process. Multiple adducts can be selected by holding down the "ctrl" button (Windows) or the "Command" button (Mac). A range of adducts can be quickly selected by clicking the first compound in the list, and while holding "Shift" select the last compound (*see* **Note 4**).

7. The "Tolerance" of the mass spectrometer is entered either as mDa (1 Da is the mass of one proton) or PPM $\left(\frac{\text{Error in } m/z}{m/z} \times 10^6\right)$ (*see* **Note 5**).
8. The results can also be filtered by the octanol–water partition coefficient (logP) or the retention index (Kovats RI) (*see* **Note 6**).
9. Halogenated compounds can be returned in the results but can be excluded if the sample analyzed is not expected to have halogenated compounds (*see* **Note** 7).
10. As before, the specific organism studied can be selected to only search compounds relevant to that organism ("Database in upper left") (*see* **Note 8**).
11. After clicking "Search," a results screen similar to Subheading 3.1, **step 9** is displayed (Fig. 2). The table of compounds has the following columns: Mine ID, *m/z* of the MINE compound, the particular adduct that matches the m/z searched, the molecular formula, the chemical structure, and the NP likeness. If logP or Kovats RI bounds were set on the prior screen, columns with these values will appear in the results screen as well.
12. Additionally, an organism or taxon of interest can be explored by typing a species name into the KEGG Genome Reconstructions box. Metabolites that are present in the corresponding KEGG metabolic model (or any metabolic model falling in the specified taxon) will be highlighted in green while compounds derived from a compound in the metabolic model will be highlighted in yellow.
13. By default, the results are sorted to present compounds from the source database (e.g., KEGG) before putative metabolites, but this may not be the correct annotation of the feature. The first row allows the user to further refine the search and resort the results by NP likeness, logP, or Kovats RI.
14. Clicking on the MINE ID brings up the web page for the compound as described in Subheading 3.1, **steps 10–12** (Fig. 3).
15. The search results can also be downloaded as a csv file, as in Subheading 3.1, **step 13**.

4 Notes

1. Searches can also be carried out using regular expressions, using the "Advanced Search" at the top of the page. This search tries to match text in the MINE database, and generally will not give as good results as the atomistic search used in "Structure Search."

2. The difference between similarity and substructure searches can be seen with the query molecule, acetate. Based on similarity, the search would return other two or three carbon organic acids, while a substructure search would return molecules like acetyl-CoA and ethyl acetate.
3. The current database only contains products one reaction away from known metabolites but the generation number can increase. For example, "2" would be a product from two promiscuous reactions in series from a known metabolite.
4. In most metabolomics protocols, $[M+H]^+$ and $[M-H]^-$ adducts are most common, although $[M+NH_4]^+$ $[M+Na]^+$ and $[M+CH_3COO]^-$ adducts may also be present depending on the salts and acids present in the sample.
5. Setting as narrow a mass window as is realistic for the analytical instrument narrows the candidate list by restricting the number of valid formulas. While not enough to unambiguously assign a correct metabolite [33], the resulting candidate lists from the MINE are more tractable for manual examination than the that of PubChem [10].
6. LogP is useful for narrowing the number of compounds when liquid chromatography is used as a separation method, while Kovats RI can be useful for limiting results in gas chromatography methods. These LogP and Kovats bounds may be set for the entire method based upon its sensitivity (e.g., filtering out high logP values in results from a hydrophilic chromatography column) or may be set for individual unknowns based on their elution order with respect to known standards.
7. Halogenation may be present in environmental samples due to metabolites of anthropogenic origin [34] (e.g., pharmaceuticals and consumer goods), but it is less common in biological samples grown in controlled laboratory conditions.
8. Use of the targeted *E. coli* or yeast databases will result in fewer potential candidates for each feature, as exogenous metabolites will not be considered. If using the KEGG MINE, a source organism can be specified with the KEGG Genomic Reconstructions to help prioritize putative metabolites produced from endogenous compounds.

Acknowledgments

National Science Foundation MCB (1614953) & MCB (1153357) and the Chicago Biomedical Consortium with support from the Searle Funds at The Chicago Community Trust.

References

1. Khersonsky O, Tawfik DS (2010) Enzyme promiscuity: a mechanistic and evolutionary perspective. Annu Rev Biochem 79:471–505
2. Humble MS, Berglund P (2011) Biocatalytic promiscuity. European J Org Chem 2011 (19):3391–3401
3. Linster CL, Van Schaftingen E, Hanson AD (2013) Metabolite damage and its repair or pre-emption. Nat Chem Biol 9(2):72–80
4. Hanson AD, Henry CS, Fiehn O, de Crécy-Lagard V (2016) Metabolite damage and metabolite damage control in plants. Annu Rev Plant Biol 67(November):131–152
5. Andersson I (2007) Catalysis and regulation in Rubisco. J Exp Bot 59(7):1555–1568
6. Rzem R, Vincent M-F, Van Schaftingen E, Veiga-da-Cunha M (2007) L-2-hydroxyglutaric aciduria, a defect of metabolite repair. J Inherit Metab Dis 30(5):681–689
7. Nam H, Lewis NE, Lerman JA, Lee D-H, Chang RL, Kim D, Palsson BO (2012) Network context and selection in the evolution to enzyme specificity. Science 337 (6098):1101–1104
8. Lerma-Ortiz C, Jeffryes JG, Cooper AJL, Niehaus TD, Thamm AMK, Frelin O, Aunins T, Fiehn O, de Crécy-Lagard V, Henry CS, Hanson AD (2016) 'Nothing of chemistry disappears in biology': the Top 30 damage-prone endogenous metabolites. Biochem Soc Trans 44(3):961–971
9. Collard F, Baldin F, Gerin I, Bolsée J, Noël G, Graff J, Veiga-da-Cunha M, Stroobant V, Vertommen D, Houddane A, Rider MH, Linster CL, Van Schaftingen E, Bommer GT (2016) A conserved phosphatase destroys toxic glycolytic side products in mammals and yeast. Nat Chem Biol 12(8):601–607
10. Jeffryes JG, Colastani RL, Elbadawi-Sidhu M, Kind T, Niehaus TD, Broadbelt LJ, Hanson AD, Fiehn O, Tyo KEJ, Henry CS (2015) MINEs: open access databases of computationally predicted enzyme promiscuity products for untargeted metabolomics. J Cheminform 7 (1):44
11. Hatzimanikatis V, Li C, Ionita JA, Henry CS, Jankowski MD, Broadbelt LJ (2005) Exploring the diversity of complex metabolic networks. Bioinformatics 21(8):1603–1609
12. Finley SD, Broadbelt LJ, Hatzimanikatis V (2009) Thermodynamic analysis of biodegradation pathways. Biotechnol Bioeng 103 (3):532–541
13. Henry CS, Broadbelt LJ, Hatzimanikatis V (2010) Discovery and analysis of novel metabolic pathways for the biosynthesis of industrial chemicals: 3-hydroxypropanoate. Biotechnol Bioeng 106(3):462–473
14. Smith CA, Want EJ, O'Maille G, Abagyan R, Siuzdak G (2006) XCMS: processing mass spectrometry data for metabolite profiling using nonlinear peak alignment, matching, and identification. Anal Chem 78(3):779–787
15. Pluskal T, Castillo S, Villar-Briones A, Oresic M (2010) MZmine 2: modular framework for processing, visualizing, and analyzing mass spectrometry-based molecular profile data. BMC Bioinformatics 11(1):395
16. Tsugawa H, Cajka T, Kind T, Ma Y, Higgins B, Ikeda K, Kanazawa M, VanderGheynst J, Fiehn O, Arita M (2015) MS-DIAL: data-independent MS/MS deconvolution for comprehensive metabolome analysis. Nat Methods 12(6):523–526
17. Keseler IM, Mackie A, Peralta-Gil M, Santos-Zavaleta A, Gama-Castro S, Bonavides-Martínez C, Fulcher C, Huerta AM, Kothari A, Krummenacker M, Latendresse M, Muñiz-Rascado L, Ong Q, Paley S, Schröder I, Shearer AG, Subhraveti P, Travers M, Weerasinghe D, Weiss V, Collado-Vides J, Gunsalus RP, Paulsen I, Karp PD (2013) EcoCyc: fusing model organism databases with systems biology. Nucleic Acids Res 41(Database issue):D605–D612
18. Jewison T, Knox C, Neveu V, Djoumbou Y, Guo AC, Lee J, Liu P, Mandal R, Krishnamurthy R, Sinelnikov I, Wilson M, Wishart DS (2012) YMDB: the yeast metabolome database. Nucleic Acids Res 40(Database issue):D815–D820
19. Kanehisa M, Goto S, Sato Y, Kawashima M, Furumichi M, Tanabe M (2014) Data, information, knowledge and principle: back to metabolism in KEGG. Nucleic Acids Res 42 (1):D199–D205
20. Flower DR (1998) On the properties of bit string-based measures of chemical similarity. J Chem Inf Model 38(3):379–386
21. Heller S, McNaught A, Stein S, Tchekhovskoi D, Pletnev I (2013) InChI—the worldwide chemical structure identifier standard. J Cheminform 5(1):7
22. Weininger D (1988) SMILES, a chemical language and information system. 1. Introduction to methodology and encoding rules. J Chem Inf Model 28(1):31–36
23. Ertl P, Roggo S, Schuffenhauer A (2008) Natural product-likeness score and its application

for prioritization of compound libraries. J Chem Inf Model 48(1):68–74

24. Stein SE, Babushok VI, Brown RL, Linstrom PJ (2007) Estimation of Kováts retention indices using group contributions. J Chem Inf Model 47(3):975–980
25. Hastings J, de Matos P, Dekker A, Ennis M, Harsha B, Kale N, Muthukrishnan V, Owen G, Turner S, Williams M, Steinbeck C (2013) The ChEBI reference database and ontology for biologically relevant chemistry: enhancements for 2013. Nucleic Acids Res 41(D1): D456–D463
26. Henry CS, DeJongh M, a Best A, Frybarger PM, Linsay B, Stevens RL (2010) High-throughput generation, optimization and analysis of genome-scale metabolic models. Nat Biotechnol 28(9):977–982
27. Velankar S, van Ginkel G, Alhroub Y, Battle GM, Berrisford JM, Conroy MJ, Dana JM, Gore SP, Gutmanas A, Haslam P, Hendrickx PMS, Lagerstedt I, Mir S, Fernandez Montecelo MA, Mukhopadhyay A, Oldfield TJ, Patwardhan A, Sanz-García E, Sen S, Slowley RA, Wainwright ME, Deshpande MS, Iudin A, Sahni G, Salavert Torres J, Hirshberg M, Mak L, Nadzirin N, Armstrong DR, Clark AR, Smart OS, Korir PK, Kleywegt GJ (2016) PDBe: improved accessibility of macromolecular structure data from PDB and EMDB. Nucleic Acids Res 44(D1):D385–D395
28. Kim S, Thiessen PA, Bolton EE, Chen J, Fu G, Gindulyte A, Han L, He J, He S, Shoemaker BA, Wang J, Yu B, Zhang J, Bryant SH (2016) PubChem substance and compound databases. Nucleic Acids Res 44(D1):D1202–D1213
29. Yasugi E, Watanabe K (2002) LIPIDBANK for web, the newly developed lipid database. Tanpakushitsu Kakusan Koso 47(7):837–841
30. Afendi FM, Okada T, Yamazaki M, Hirai-Morita A, Nakamura Y, Nakamura K, Ikeda S, Takahashi H, Altaf-Ul-Amin M, Darusman LK, Saito K, Kanaya S (2012) KNApSAcK family databases: integrated metabolite-plant species databases for multifaceted plant research. Plant Cell Physiol 53(2):e1
31. Wishart DS, Jewison T, Guo AC, Wilson M, Knox C, Liu Y, Djoumbou Y, Mandal R, Aziat F, Dong E, Bouatra S, Sinelnikov I, Arndt D, Xia J, Liu P, Yallou F, Bjorndahl T, Perez-Pineiro R, Eisner R, Allen F, Neveu V, Greiner R, Scalbert A (2013) HMDB 3.0—the human metabolome database in 2013. Nucleic Acids Res 41(Database issue):D801–D807
32. Fahy E, Subramaniam S, Murphy RC, Nishijima M, Raetz CRH, Shimizu T, Spener F, van Meer G, Wakelam MJO, Dennis EA (2009) Update of the LIPID MAPS comprehensive classification system for lipids. J Lipid Res 50(Suppl):S9–S14
33. Kind T, Fiehn O (2006) Metabolomic database annotations via query of elemental compositions: mass accuracy is insufficient even at less than 1 ppm. BMC Bioinformatics 7(1):234
34. Bletsou AA, Jeon J, Hollender J, Archontaki E, Thomaidis NS (2015) Targeted and non-targeted liquid chromatography-mass spectrometric workflows for identification of transformation products of emerging pollutants in the aquatic environment. TrAC Trends Anal Chem 66:32–44

Chapter 3

Antisense RNA Elements for Downregulating Expression

Yaping Yang, Jian Wang, Ruihua Zhang, and Yajun Yan

Abstract

Antisense RNA (asRNA) technology is an important tool for downregulating gene expression. When applying this strategy, the asRNA interference efficiency is determined by several elements including scaffold design, loop size, and relative abundance. Here, we take the *Escherichia coli* gene *fabD* encoding malonyl-CoA-[acyl-carrier-protein] transacylase as an example to describe the asRNA design with reliable and controllable interference efficiency. Real-time PCR and fluorescence assay methods are introduced to detect the interference efficiency at RNA level and protein level, respectively.

Key words asRNA, Scaffold design, Loop size, Relative abundance, Interference efficiency detection, Gene *fabD*

1 Introduction

Antisense RNA (asRNA)-mediated gene silencing has been extensively utilized as a critical tool to control gene expression without any genomic modification of the target gene [1–3]. Through base pairing between designed asRNA and target mRNA, the resulting double-strand RNA (dsRNA) can inhibit translation initiation by steric hindrance and trigger degradation of the target mRNA by dsRNA-specific RNases (*see* Fig. 1c) [4–6]. There are advantages of employing the asRNA strategy. Unlike the traditional gene knockout strategy for disrupting gene expression, the asRNA triggered gene downregulation still maintains the cell viability. The plasmids for asRNA expression can be easily constructed, conditionally controlled by various promoters, and designed to target several genes simultaneously. However, the asRNA-mediated gene inhibition requires continuous presence of abundant asRNA molecules. Thus, the reliable supply and stability of asRNA are prerequisite for this strategy to be applied efficiently in genetic manipulations. The asRNA interference efficiency is mainly affected by several factors, among which asRNA stability is the primary player [7, 8]. Previous reports have shown that asRNA scaffold was

Christine Nicole S. Santos and Parayil Kumaran Ajikumar (eds.), *Microbial Metabolic Engineering: Methods and Protocols*, Methods in Molecular Biology, vol. 1927, https://doi.org/10.1007/978-1-4939-9142-6_3,

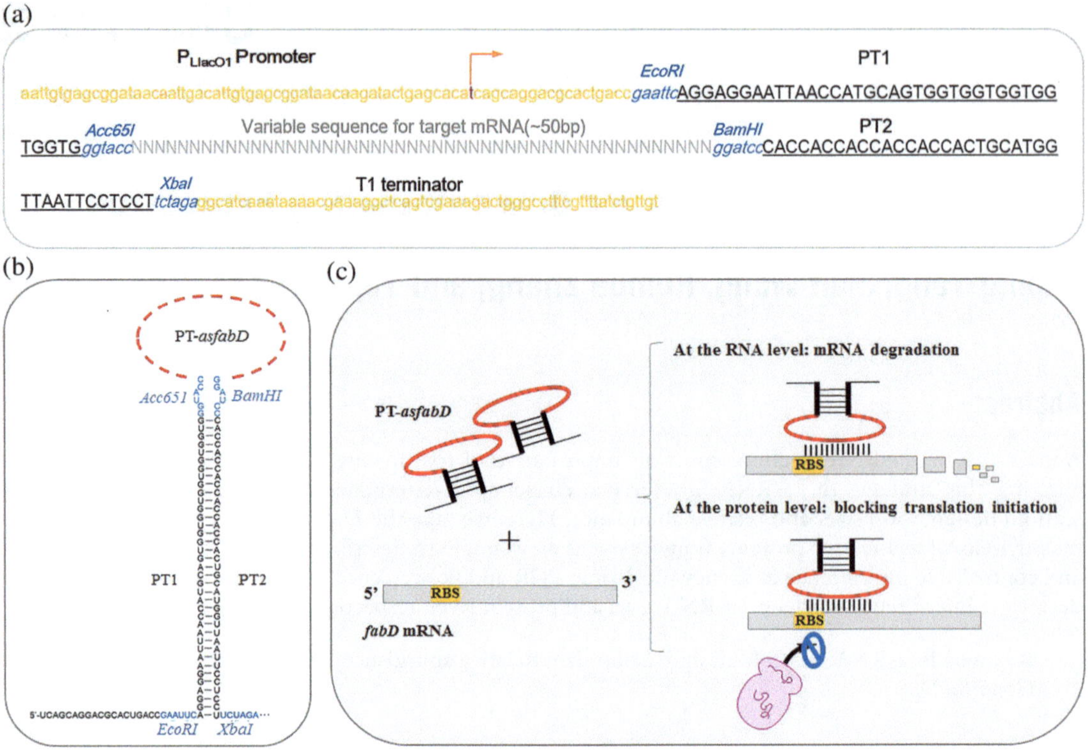

Fig. 1 The stem-loop scaffold and asRNA interference mechanism. (**a**) The DNA sequence of two 38 bp inverted repeats. (**b**) *asfabD* scaffold. When *asfabD* is transcribed in *E. coli*, the two PT would complementarily pair with each other and generate a double-strand stem structure, meanwhile the interposed *asfabD* sequence would form the loop. (**c**) asRNA interference mechanism. When asRNA complementarily binds to target mRNA, the resulting double strand RNA (dsRNA) will trigger degradation of the target mRNA or block translation initiation by steric hindrance

beneficial to asRNA stability, like Hfq protein-assisted asRNA scaffold showed an increased lifetime and effectiveness in target binding [9, 10]. A stem-loop asRNA scaffold also exhibited a prolonged half-life and efficient target recognition without the involvement of any protein [11] (*see* Fig. 1). For this stem-loop scaffold, two 38 bp inverted repeats were the essential parts to maintain the asRNA stability [12] (*see* Fig. 1(b)). Herein, we create an efficient method to assemble the basic stem-loop scaffold with a random DNA fragment. By replacing the random DNA fragment with a specific sequence complementary to the translation initiation region of any target gene, various new artificial regulatory asRNAs would be created [13]. Another strategy to improve asRNA interference efficiency is to optimize the asRNA loop size. Our previous work reported that interference efficiency was negatively correlated with the asRNA loop size and asRNA with 100-nt loop length conducted up to 80% interference efficiency [13]. To simplify the experimental operation as well as to ensure high interference efficiency, we recommend 100-nt as the primary choice for the asRNA loop design [13]. Finally, the relative asRNA abundance is crucial to enhance interference efficiency [11, 14]. The excess intracellular asRNAs enable an efficient competition over ribosome to sequester target mRNAs. Thus, we recommend constructing asRNAs on the high copy number plasmid or under control of strong promoters. Real-time quantitative PCR and eGFP fluorescence assay are performed to detect the interference efficiency at RNA level and protein level. Here, by taking *Escherichia coli* malonyl-CoA-[acyl-carrier-protein] transacylase gene *fabD* as an example, we demonstrate the asRNA design with reliable and controllable interference efficiency.

2 Materials

Prepare all solutions by deionized water (diH_2O) and analytical grade reagents. Autoclave all solutions at 121 °C for 20 min. Prepare and store all reagents at room temperature (unless specified otherwise).

2.1 Construction of asRNA Interference Plasmids

1. Backbone plasmids: high copy number plasmid pZE12-luc and low copy number plasmid pSA74.
2. Polymerase chain reaction (PCR): Genomic DNA of *E. coli* BW25113, primers (*see* Table 1), Phusion High-Fidelity DNA polymerase, 10 × buffer, DMSO (if necessary).
3. Restriction endonucleases: *EcoR*I, *Acc65*I, *BamH*I, *Xba*I, *Sph*I, *Spe*I.
4. Tris–acetate–EDTA (TAE) Buffer (50×): 242 g Tris free base, 18.61 g disodium EDTA, 57.1 ml glacial acetic acid, and diH_2O to a final volume of 1 l.
5. 0.8% agarose gel: 0.8 g agarose in 100 ml TAE buffer.
6. DNA ladder: GeneRuler 1 kb DNA ladder.
7. All kits for gene cloning: PCR purification kit, Gel DNA recovery kit, Quick ligation kit, Plasmid miniprep kit.
8. Competent cells: *E. coli* XL-Blue and BW25113.
9. LB medium: 10 g tryptone, 5 g yeast extract, and 10 g NaCl in diH_2O to a final volume of 1 liter.
10. Agar plates: each 9 cm diameter petri dish contains 10 ml LB medium containing 1.5% agar.
11. Antibiotics: ampicillin (100 mg/l), chloramphenicol (34 mg/l).

2.2 Fluorescence Assay

1. eGFP cDNA.
2. Inducer: 0.5 M isopropyl β-D-1-thiogalactopyranoside (IPTG).

Table 1
Primer sequences for cloning and qRT-PCR analysis

Primers	Primer Sequence (5′–3′)
1	AAAAAgaattcAggaggaaTTaaccaTgcagTggTggTggTggTggTgGGTACCgatatc
2	aaaaaGcatgccatggcggccgcaagcttgatatcGGTACCcAccAccAccAccAccAc
3	AAAAAgcatgCgatatcaagcttgcggccgccatggatccCaccaccaccaccaccacTg
4	AAAAAtctagaAggAggAAttAAccAtgcAgtggtggtggtggtggtGggatccatggcg
5	AAAAAGGTACCCCGCCATATCAGCCAGCATTCCAAC
6	AAAAAGGATCCAGTTGGTTCTGCTTGAAGCCTTTGG
7	GCTCGTGTTGTGAAATGTT
8	TGTAGCCCTGGTCGTAAGG
9	TGAAGAACTGAATAAAACC
10	GCAATAGACGCATCATCCA
11	aaaaaggtaccAACTCGGTATGTCTATGGATAATGTC
12	ACAGCTCCTCGCCCTTGCTCACAAGCTCGAGCGCCGCTGCCATC
13	GATGGCAGCGGCGCTCGAGCTTGTGAGCAAGGGCGAGGAGCTGT
14	aaaaaggatccTTACTTGTACAGCTCGTCCATGCCGAG

3. Black 96-well plate with clear bottom.
4. BioTek plate reader.

2.3 Quantitative Real-Time (qRT) PCR Analysis

All materials should be RNase-free in this section. All solutions are prepared by RNase-free water and containers.

1. Total RNA isolation: RNA purification kit, RNase-free pipette tips, RNase free water, 1.5 ml RNase-free microcentrifuge tubes, micro tube homogenizer system, DNase I, 2-mercaptoethanol, 100% ethanol, 10% SDS (0.5 μl/sample), Lysozyme solution (100 μl/sample): 10 mM Tris–HCl (pH 8.0), 0.1 mM EDTA, 10 mg/ml lysozyme, Lysis buffer: before beginning of the lysis and homogenization steps, prepare a fresh amount of lysis buffer (10 μl 2-mercaptoethanol/1 ml lysis buffer). Use 350 μl fresh lysis buffer for 1×10^9 bacterial cell.
2. cDNA synthesis: Primers (*see* Table 1), Protoscript II first strand cDNA synthesis kit, sterile aerosol pipette tips, RNase-free water.
3. qRT-PCR: primers (*see* Table 1), FastStart Universal SYBR Green Master, 8 strip PCR tube.

4. Required equipment: NanoDrop 2000c UV-Vis Spectrophotometer, iQ™ SYBR Green Supermix.
5. M9Y medium: 10 g glucose, 6 g Na_2HPO_4, 0.5 g NaCl, 3 g KH_2PO_4, 1 g NH_4Cl, 1 mM $MgSO_4$ (final concentration), 0.1 mM $CaCl_2$ (final concentration), 2 g yeast extract, 2 g sodium citrate, and 100 mg serine dissolved in diH_2O with a final volume of 1 liter.

3 Methods

3.1 Construction of the Stem-Loop Scaffold Carrying asRNA

3.1.1 Construct Parent Plasmid pZE-PT (See Fig. 2, Note 1)

1. Amplify the DNA fragment 1 by PCR reaction. Dilute primers 1 and 2 using diH_2O to 1 μM. Prepare the PCR mixture (50 μl) using Phusion High-Fidelity DNA polymerase as follows: diH_2O: 30.5 μl, 10 × buffer: 10 μl, 2 mM dNTP: 5 μl, primers 1 and 2: 2 μl (each), Enzyme: 0.5 μl. During this mixture, primers 1 and 2 serve as both primers and template. The PCR product of DNA fragment 1 is 85 bp (*see* **Note 2**). The PCR program is: predenaturation at 98 °C for 3 min, 30 cycles of 98 °C for 30 s, 60 °C for 30 s and 72 °C for 20 s, then extension at 72 °C for another 5 min, followed by cooling down to 16 °C.
2. Amplify the DNA fragment 2 by PCR reaction using primers 3 and 4. The procedures are the same as Subheading 3.1.1, **step 1**. The PCR product of DNA fragment 2 is 85 bp (*see* **Note 2**).
3. Purify DNA fragments 1 and 2 using PCR purification kit. Add 100 μl binding buffer into each 50 μl DNA fragment. Mix by pipette. Transfer the mixture into a spin column with a collection tube. After 1 min, centrifuge the spin column at 6000 × *g* for 1 min. Add 200 μl wash buffer into the spin column. Then centrifuge at 13,500 × *g* for 30 s. Transfer column to a 1.5 ml centrifuge tube. After 1 min, add 30 μl diH_2O to elute the DNA from the spin column (*see* **Note 3**).
4. Prepare a mixture of enzyme digestion as follows. For DNA fragments 1 and 2: diH_2O: 17 μl, DNA: 15 μl, 10 × buffer: 4 μl, enzymes (*see* **Note 4**): 2 μl (each), total volume: 40 μl. For plasmid pZE-luc: diH_2O: 26 μl, DNA: 6 μl, Buffer: 4 μl, Enzymes (*see* **Note 4**): 2 μl (each), total volume: 40 μl. Incubate the digested mixture at 37 °C for 2–4 h in a water bath.
5. Purify the digested products. For DNA fragments 1 and 2: we use the same method as Subheading 3.1.1, **step 3** and 6 μl diH_2O is added for elution. For plasmid pZE-luc: load digested product into a 0.8% agarose gel. Run the gel in the 1 × TAE buffer using 100 constant voltage for 20 min. Observe the gel under low UV. Cut the gel containing target fragment, and place it in a centrifuge tube and purified the DNA using gel DNA recovery kit. Add ADB buffer (300 μl/100 mg, the

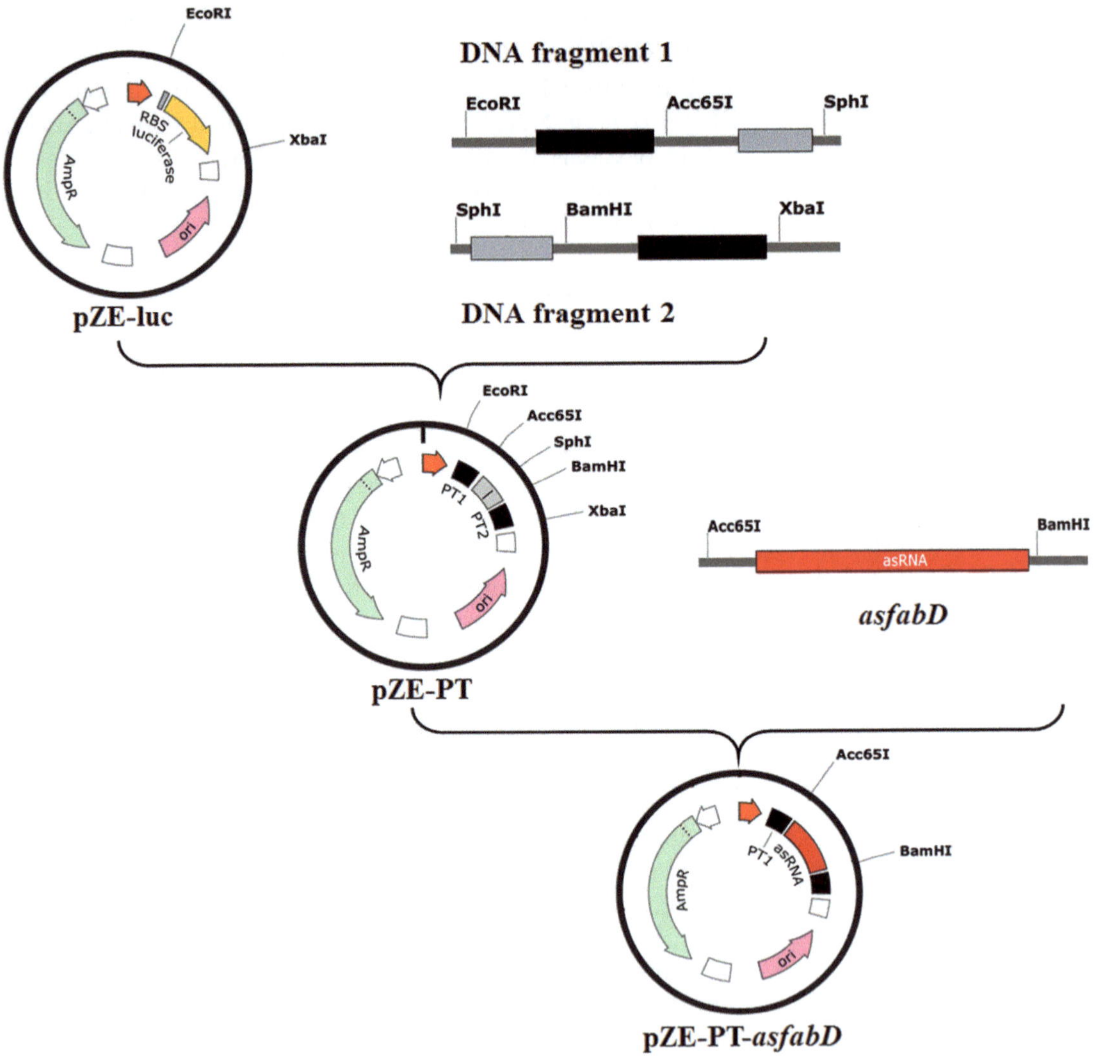

Fig. 2 The plasmid construction of pZE-PT-*asfabD.* DNA fragment 1 and DNA fragment 2 were inserted into pZE-luc in between *EcoRI* and *SphI*, *SphI* and *XbaI*, respectively, to generate parent plasmid pZE-PT. The DNA sequence of *asfabD* was inserted into plasmid pZE-PT in between *Acc65I* and *BamHI* to produce pZE-PT-*asfabD*

volume of buffer over the weight of agarose) into the centrifuge tube. Melt the gel at 55 °C for 5 min in a water bath. After the gel completely melts, place it in a spin column with a collection tube. The next step is the same as Subheading 3.1.1, **step 3**. But we add 6 μl diH_2O for elution.

6. Prepare a ligation mixture. DNA fragments 1 and 2: 5 μl (each), plasmid pZE-luc: 2 μl, 2× ligation buffer: 13 μl, ligase: 1 μl, total volume: 26 μl. Wait for 20 min at room temperature and then proceed for transformation.

7. Transfer all the reaction mixture of ligation into 50 μl chemical competent cell *E. coli* XL-Blue. Keep on ice for 20 min. Then heat shock at 42 °C for 90 s. Keep on ice for 2 min. Add 1 ml LB medium into the mixture.
8. Put the culture into the shaker for 40 min at 37 °C. Centrifuge the culture at 1500 × *g* for 3 min. Discard the supernatant and resuspend the cell pellet in 100 μl LB medium. Spread them on the LB agar plate containing 100 mg/l ampicillin. Incubate the plate overnight at 37 °C.
9. Pick three transformants and incubate them into 3 ml LB liquid broth at 37 °C for 12 h.
10. Collect the cell by centrifuge at 13,500 × *g* for 1 min. Extract the plasmid by plasmid miniprep kit. Add 50 μl diH_2O to elute the plasmid DNA.
11. Verify the plasmids by *Spe*I and *Sph*I. Prepare mixture of enzyme digestion as follows: diH_2O: 6 μl, DNA: 2 μl, 10 × buffer: 1 μl, enzymes: 0.5 μl (each), total volume: 10 μl (*see* **Note 5**).

*3.1.2 Construct Target Plasmid pZE-PT-*asfabD *as Follows (See* **Note 6***)*

1. Amplify the gene *asfabD* by PCR reaction. Dilute primers 5 and 6 using diH_2O to 1 μM. Prepare the PCR mixture (50 μl) using Phusion High-Fidelity DNA polymerase as follows: diH_2O: 31.5 μl, 10 × buffer: 10 μl, 2 mM dNTP: 5 μl, primers 5 and 6: 1 μl (each), genomic DNA: 1 μl, enzyme: 0.5 μl. The PCR product is 100 nt (*see* **Note 7**).
2. Purify the DNA product using the same method as Subheading 3.1.1, **step 3**. Add 30 μl diH_2O for elution.
3. Prepare digestion mixture. For DNA fragment as*fabD*: diH_2O: 17 μl, DNA: 15 μl, 10 × buffer: 4 μl, enzymes: 2 μl (each), total volume: 40 μl. For parent plasmid pZE-PT: diH_2O: 26 μl, DNA: 6 μl, 10 × buffer: 4 μl, enzymes: 2 μl (each), total volume: 40 μl. Incubate the digested mixture at 37 °C for 2–4 h in a water bath (*see* **Note 8**).
4. Purify the digested product. For DNA fragment as*fabD*: we use the same method as Subheading 3.1.1, **step 3**. But, we add 6 μl diH_2O for elution. For parent plasmid pZE-PT: we used the same method as Subheading 3.1.1, **step 5**. We add 6 μl diH_2O for elution.
5. Prepare ligation mixture. DNA fragment as*fabD*: 4 μl, plasmid pZE-PT: 1 μl, 2 × ligation buffer: 5.5 μl, Quick DNA ligase: 0.5 μl, total volume: 11 μl. Place the reaction tube at room temperature for 20 min.
6. Heat-shock transformation and plasmid verification of pZE-PT-*asfabD*. We used the same method from Subheading 3.1.1, **steps 7–11**.

3.2 Interference Efficiency Detection

3.2.1 Direct Detection Method-qRT PCR (See **Note 9**)

1. Transform the plasmid pZE-PT-*asfabD* into the target strain *E. coli* BW25113. Transform the parent plasmid pZE-PT into the target strain *E. coli* BW25113 as a negative control (*see* **Note 10**). Pick three transformants of pZE-PT-*asfabD* and pZE-PT from the overnight plates, respectively. Incubate these transformants in 3.5 ml LB medium containing ampicillin (100 mg/l) at 37 °C and 290 rpm for 5 h.
2. Transfer 2% cultures into baffled flasks containing 15 ml of M9Y media with 1 mM IPTG at 30 °C and 290 rpm.
3. After 12 h, take 500 μl cell cultures and centrifuge at 13,500 × *g* for 1 min at 4 °C to pellet cells. Discard the supernatant (*see* **Note 11**).

Total RNA purification process:

4. Add 100 μl lysozyme solution (*see* Subheading 2.3, **item 1**) to the cell pellet and resuspend by vortexing.
5. Add 0.5 μl 10% SDS solution. Vortex to mix well.
6. Wait for 5 min at room temperature. Add 350 μl lysis bufer with prepared 2-mercaptoethanol (*see* Subheading 2.3, **item 1**). Vortex to mix well.
7. Homogenize the cell lysate: transfer the lysate to a 1.5 ml RNase-free microcentrifuge tube and homogenize using an micro tube homogenizer system with sterile micropestle for at least 45 s. Centrifuge the homogenate at 850 × *g* for 5 min at room temperature, then transfer the supernatant to clean RNase-free microcentrifuge tube.
8. Add 250 μl 100% ethanol to each volume of bacterial cell homogenate. Vortex to mix completely to disperse any visible precipitate that may form after adding ethanol.
9. Transfer the sample (including any remaining precipitate) to a spin column with a collection tube. Centrifuge at 13,500 × *g* for 15 s at room temperature. Descard the flow-trough.
10. Prepare DNase I reaction mixture for on-column treatment as follows: 10 × DNase I reaction buffer: 8 μl, resuspended DNase I (3 U/μl): 10 μl, RNase-free water: 62 μl, final volume: 80 μl. Add the 80 μl DNase mixture directly onto the surface of the spin column. Incubate at room temperature for 15 min.
11. Add 350 μl Wash buffer I to the spin column. Centrifuge at 13,500 × *g* for 15 s at room temperature. Discard the flow-through.
12. Add 500 μl Wash buffer II with ethanol to the spin column. Centrifuge at 13,500 × *g* for 15 s at room temperature. Discard the flow-through. Repeat this step.

13. Centrifuge the spin column at 13,500 × *g* for 1 min. Insert the spin column into a new collection tube. Add 30 μl RNase-free water into the center of the spin column.

14. Incubate at room temperature for 1 min. Centrifuge spin column and collection tube at 13,500 × *g* and room temperature for 1 min.

15. Analyzing RNA concentration and quality by NanoDrop (*see* **Note 12**).

First Strand cDNA synthesis:

16. Dilute the denatured RNA to 600 ng/μl. Mix RNA sample and primer in a sterile RNase-free microfuge tube as follows: Template RNA: 1 μl, $d(T)_{23}VN$ (50 μM):2 μl, RNase-free H_2O: 5 μl. Total volume is 8 μl. Denature sample RNA and primer d $(T)_{23}VN$ at 65 °C for 5 min. Spin briefly and put on ice imediately.

17. Add the following components: 2 × Protoscript II reaction mix: 10 μl, 10× Protoscript II enzyme mix: 2 μl. Incubate the 20 μl cDNA synthesis reaction at 42 °C for 1 h. Inactivate the enzyme at 80 °C for 5 min (*see* **Note 13**).

qRT-PCR reaction:

18. Dilute the cDNA to 12 ng/μl. Dilute the primers to 30 μM. In a 1.5 ml reaction tube on ice, prepare the PCR mixture as follows: FastStart SYBR Green Master (ROX): 25 μl, primers (30 μM): 0.5 μl (each), diH_2O: 22 μl. Total volume is 45 μl. Mix the solutions carefully by pipetting it up and down. Do not vortex.

19. Pipet 24 μl PCR mix into each PCR reaction vessel. Add 1 μl of template DNA. Mix carefully by pipetting up and down. Seal the tubes with clear tube caps. Keep triplicate for each sample. We choose 16S RNA as a normalizer gene from *E. coli*. By using the primers 7 and 8, the product size of 16S RNA is 150 bp. The product size of target gene *fabD* is 274 bp when using primers 9 and 10 (*see* **Note 14**). Set the cycling conditions as follows: 95 °C for 30 s; 95 °C for 10 min; followed by 45 cycles of 95 °C for 15 s; 58 °C for 30 s; 72 °C for 20 s (*see* **Note 15**).

20. Conduct a melting curve analysis for each reaction using the following condition: 55 °C to 95 °C; 0.5 °C/read; 1 s hold; 72 °C for 10 min (*see* **Note 16**).

21. Calculate the relative gene expression by the $2^{-\Delta\Delta Ct}$ approximation method (*see* **Note 17**).

3.2.2 Indirect Detection Method—Fluorescence Assay (See **Note 18**)

Construct plasmid pSA-*pfabD/egfp* (*see* **Note 19**):

1. Amplify gene *pfabD* and gene *egfp* by PCR reaction. All procedures are the same as Subheading 3.1.1, **step 12**. Use primers 11 and 12 to amplify gene *pfabD*. Use primers 13 and 14 to amplify *egfp*. Purify the PCR product by gel recovery kit. The procedures are the same as Subheading 3.1.1, **step 5**. Add 40 μl diH_2O for elution (*see* **Note 19**).
2. Overlap extension PCR for *pfabD/egfp*. Dilute PCR product of *pfabD* and *egfp* by fivefolds, respectively. Prepare the PCR mixture (50 μl) using Phusion High-Fidelity DNA polymerase as follows: diH_2O: 31.5 μl, 10× buffer: 10 μl, 2 mM dNTP: 5 μl, primers 11 and 14: 1 μl (each), diluted *pfabD* and *egfp*: 1 μl (each), enzyme: 0.5 μl. Purify the PCR product by gel recovery kit. The procedures are the same as Subheading 3.1.1, **step 5**. Add 30 μl diH_2O for elution from the spin column.
3. Prepare digestion mixture. For DNA fragment *pfabD/egfp*: diH_2O: 22 μl, DNA: 10 μl, 10 × buffer: 4 μl, enzymes: 2 μl (each) (*see* **Note 20**), total volume: 40 μl. For plasmid pSA74: diH_2O: 26 μl, DNA: 6 μl, 10 × buffer: 4 μl, Enzymes: 2 μl (each), total volume: 40 μl. Incubate the digested mixture at 37 °C for 2–4 h in a water bath.
4. The following cloning procedures are the same from Subheading 3.1.1, **steps 5–11**.

   Fluorescence Assay.
5. Cotransform two plasmids pZE-PT-*asfabD* and pSA-*pfabD/egfp* into target strain *E. coli* BW25113. Cotransform the parent plasmid pZE-PT with pSA-*pfabD/egfp* into *E. coli* BW25113 as a negative control (*see* **Note 21**).

   Inoculate three transformants of each strain into 3.5 ml LB medium with ampicillin (100 mg/l) and chloramphenicol (34 mg/l), respectively at 37 °C and 290 rpm for 5 h.
6. Transfer 2% cultures into a black 96-well plate with clear bottom containing 200 μl LB medium with 1 mM IPTG. eGFP fluorescence is monitored by spectrometer (Biotek) with a time duration of 24 h. Set the conditions as follows: Temperature: Setpoint 37 °C; Start Kinetic [Run 24:00:00, Interval 0:20:00], Shake: Medium for 16:30, Read: (A) 600, Read: (F) 485/20 nm, 528/20 nm; End Kinetic. Set the reading type as endpoint mode (*see* **Note 22**).
7. Subtract the background cell fluorescence of each sample and normalize the eGFP fluorescence intensity of each sample against its OD_{600}.

4 Notes

1. To keep a high abundance of asRNAs, we utilized high copy number plasmid pZE-luc for the construction of asRNA plasmid. Parent plasmid pZE-PT consisting of a random DNA sequence in between the two PT sequences, which serves as the basis for constructing various asRNA scaffold plasmids (*see* Fig. 1(a)).
2. Since the inverted repeat sequences are consisted of GC-rich nucleotides, the whole PT sequence is hard to be amplified by overlap extension PCR. Thus, we divided the whole PT sequence into two part including DNA fragment 1 and DNA fragment 2 (*see* Fig. 2).
3. Since the size of DNA fragments 1 and 2 is small, we use 30 μl diH_2O to elute DNA product to increase the recovery efficiency.
4. We use enzymes *EcoR*I and *Sph*I to digest DNA fragment 1 and use enzymes *Sph*I and *Xba*I to digest DNA fragment 2. For the pZE-luc plasmid, we use enzymes *EcoR*I and *Xba*I for the digestion.
5. When verifing the plasmid, we recommend that the two chosen enzyme restriction sites are located in plasmid and the target DNA, respectively.
6. By replacing target-binding sequence, we can generate various new synthetic asRNAs.
7. Primer 5 is reversely complementary to the target gene *fabD* starting from the +30 nt coding region. Primer 6 is identical to the target gene *fabD* starting from −70 nt ahead of the coding region.
8. We use enzymes *Acc65*I and *BamH*I to digest *asfabD* fragment and plasmid pZE-PT.
9. Since asRNA triggered mRNA degradation is the main mechanism of asRNA inhibition, the real-time PCR is the direct method for the identification of interference efficiency at the RNA level.
10. The strain containing plasmid pZE-PT would show higher mRNA concentration of *fabD* than the strain containing plasmid pZE-PT-*asfabD*.
11. For better results, the cells in log-phase growth are optimal to isolate RNA. To prevent RNA degradation, it is important to disrupt the cells fast and completely.
12. Before each experiment, clean your bench and relative equipment by 70% ethanol. During total RNA isolation process, please always wear a mask and frequently replace gloves to avoid RNA

contamination or degradation. The ratio of $OD_{260/280}$ is used to measure RNA purity. An $OD_{260/280} > 1.8$ indicates that RNA is reasonably clean of proteins and other UV chromophores that could negatively affect the stability of the stored RNA.

13. After quantifying the total RNA concentration, we dilute each sample to the same concentration. Then, we take the same volume of total RNA as the template for the cDNA synthesis in the next step. Heat denaturation of the total RNA is required to open the secondary structure which may impede long cDNA synthesis.
14. When designing real-time PCR primers for target genes, it is necessary to confirm the primer specificity by the BLAST analysis against the NCBI genome database. To avoid nonspecific bands of the target gene, at least one designed primer should be uniquely homologous to the genomic DNA.
15. It is optional to verify the integrities of the obtained products by gel electrophoresis.
16. Melting curve analysis is necessary and important to assess the specificity of the real-time PCR products. A bipeak curve indicates a contamination from genomic DNA. A peak in the blank control indicated a contamination from the diH_2O or reagent.
17. The reference introduced a detailed calculation process as an example [15].
18. Fluorescence assay is an alternative detection method of asRNA interference efficiency to save experimental cost. By constructing a *pfabD/egfp* fusion protein with a low copy number plasmid, we detect the expression level of gene *fabD* by monitoring fluorescence signal. However, since the fusion protein is constructed in a plasmid, it is not a direct method to detect asRNA interference efficiency.
19. To maximally simulate the expression of chromosomal genes, we keep the native promoter for gene *fabD* and termed *pfabD*. Remove stop codon of gene *pfabD* and start codon of gene *egfp* when designing the overlap extension PCR primers.
20. We use enzymes *Acc65*I and *BamH*I to digest *pfabD/egfp* fragment and plasmid pSA74.
21. The strain containing parent plasmid pZE-PT and pSA-*pfabD/egfp* would show higher fluorescence intensity than the strain containing asRNA plasmid pZE-PT-*asfabD* and pSA-*pfabD/egfp*.
22. The procedures were applied for real-time monitoring of fluorescence intensity and OD_{600} by Gene5 software in BioTek plate reader.

References

1. Weintraub H, Izant JG, Harland RM (1985) Anti-sense Rna as a molecular tool for genetic-analysis. Trends Genet 1:22–25
2. Chen GZ, Patten CL, Schellhorn HE (2003) Controlled expression of an rpoS antisense RNA can inhibit RpoS function in Escherichia coli. Antimicrob Agents Ch 47:3485–3493
3. Yin DZ, Ji YD (2002) Genomic analysis using conditional phenotypes generated by antisense RNA. Curr Opin Microbiol 5:330–333
4. Ji YD, Yin DZ, Fox B, Holmes DJ, Payne D, Rosenberg M (2004) Validation of antibacterial mechanism of action using regulated antisense RNA expression in Staphylococcus aureus. FEMS Microbiol Lett 231:177–184
5. Rasmussen LCV, Sperling-Petersen HU, Mortensen KK (2007) Hitting bacteria at the heart of the central dogma: sequence-specific inhibition. Microb Cell Factories 6
6. Carpousis AJ (2003) Degradation of targeted mRNAs in Escherichia coli: regulation by a small antisense RNA. Genes Dev 17:2351–2355
7. Wagner EGH, Altuvia S, Romby P (2002) Antisense RNAs in bacteria and their genetic elements. Adv Genet 46:361–398
8. Tilley LD, Hine OS, Kellogg JA, Hassinger JN, Weller DD, Iversen PL, Geller BL (2006) Gene-specific effects of antisense phosphorodiamidate morpholino oligomer-peptide conjugates on Escherichia coli and Salmonella enterica serovar typhimurium in pure culture and in tissue culture. Antimicrob Agents Ch. 50:2789–2796
9. Ross JA, Ellis MJ, Hossain S, Haniford DB (2013) Hfq restructures RNA-IN and RNA-OUT and facilitates antisense pairing in the Tn10/IS10 system. RNA 19:670–684
10. Na D, Yoo SM, Chung H, Park H, Park JH, Lee SY (2013) Metabolic engineering of Escherichia coli using synthetic small regulatory RNAs. Nat Biotechnol 31:170–174
11. Nakashima N, Tamura T, Good L (2006) Paired termini stabilize antisense RNAs and enhance conditional gene silencing in Escherichia coli. Nucleic Acids Res 34:e138
12. Nakashimaand N, Tamura T (2009) Conditional gene silencing of multiple genes with antisense RNAs and generation of a mutator strain of Escherichia coli. Nucleic Acids Res 37: e103
13. Yang YP, Lin YH, Li LY, Linhardt RJ, Yan YJ (2015) Regulating malonyl-CoA metabolism via synthetic antisense RNAs for enhanced biosynthesis of natural products. Metab Eng 29:217–226
14. Cruzand C, Houseley J (2014) Endogenous RNA interference is driven by copy number. Elife 3
15. Livakand KJ, Schmittgen TD (2001) Analysis of relative gene expression data using real-time quantitative PCR and the 2(T)(−Delta Delta C) method. Methods 25:402–408

Chapter 4

Chemically Inducible Chromosomal Evolution (CIChE) for Multicopy Metabolic Pathway Engineering

Aaron M. Love, Bradley W. Biggs, Keith E. J. Tyo, and Parayil Kumaran Ajikumar

Abstract

Chemically inducible chromosomal evolution (CIChE) was developed for stable multicopy chromosomal integration of heterologous genes. In this technique, flanking an antibiotic selection marker and a gene of interest with identical regions of homology permits gene duplication via *recA* mediated homologous recombination. A strong selective pressure for gene duplication can be applied by increasing antibiotic concentration, and in a week's time one can create a set of strains with a wide range of cassette copy numbers (upward of 20×), which can be made stable by deletion of *recA*. Herein, we describe a generalized workflow for this methodology.

Key words Chromosomal integration, Molecular cloning, Metabolic engineering, Synthetic biology, *Escherichia coli*, Tools

1 Introduction

In metabolic engineering, tuning the expression of pathway enzymes is often necessary to optimize the production of a molecule of interest [1]. Typically, this is accomplished by varying promoter strength and copy number using plasmid-based systems, due to their ease of manipulation. However, plasmid-based systems are known to possess limitations, including poor long-term stability, especially under heavy metabolic burden [2]. Although this is generally tolerable during initial strain construction and screening, industrial production strains ought to be more robust. One common solution is to chromosomally integrate a pathway, which forces ordered inheritance. While methods continue to be developed to this end, they generally remain more laborious than

---

Aaron M. Love and Bradley W. Biggs contributed equally to this work.

Christine Nicole S. Santos and Parayil Kumaran Ajikumar (eds.), *Microbial Metabolic Engineering: Methods and Protocols*, Methods in Molecular Biology, vol. 1927, https://doi.org/10.1007/978-1-4939-9142-6_4,

plasmid-based systems, restricting the user to only a handful of integrations. This may preclude achieving optimal expression levels.

Chemically inducible chromosomal evolution (CIChE), in contrast, provides a facile and rapid (just over a week) approach for preparing a strain with a wide range of chromosomal copy numbers (upward of 20×). This is accomplished by first integrating a cassette ready for tandem gene duplication (TGD), which consists of a gene of interest and antibiotic selection marker that are flanked by identical homologous regions. Once the cassette is integrated, the strain is subcultured into media with increasing concentrations of antibiotic, thus increasing the selective pressure for gene duplication. This process proceeds through *recA* mediated homologous recombination to produce head-to-tail uneven tandem gene duplication events [3, 4] (Fig. 1). Eventually, CIChE duplicates the

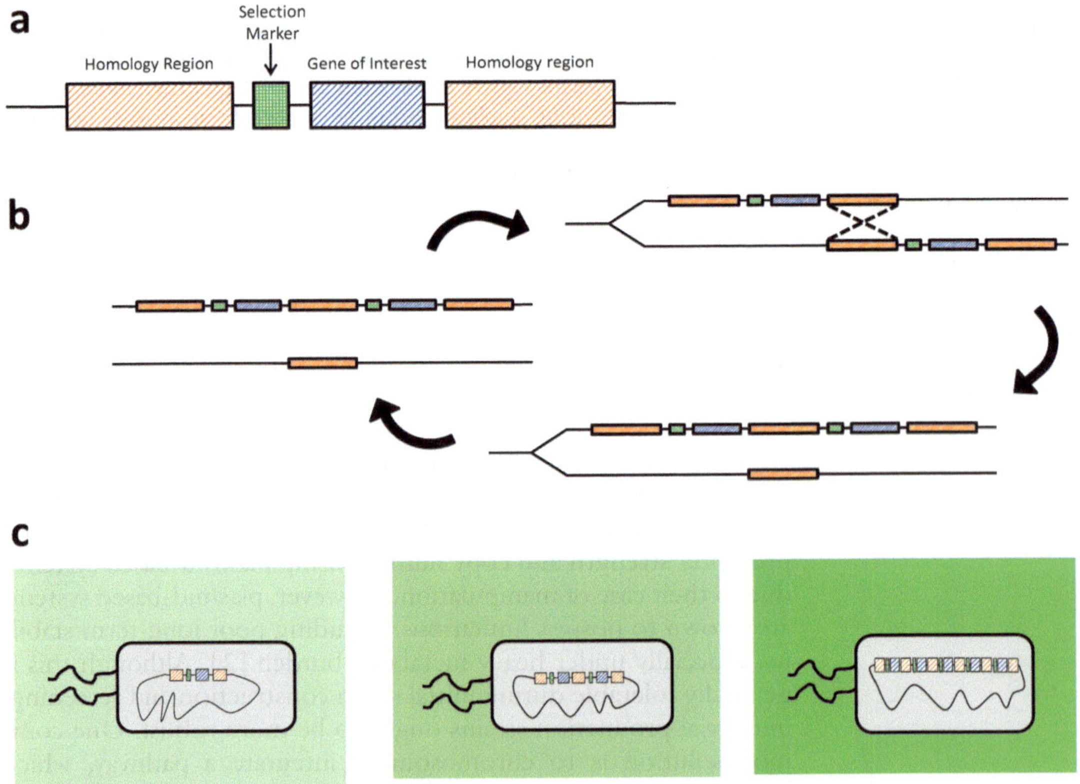

Fig. 1 Design and process of chemically inducible chromosomal evolution (CIChE). (**a**) Schematic for duplication cassette. Each cassette contains two identical ~1 kb flanking regions with little homology to the host organism. Inside the flanking region are both the gene of interest and a selection marker (i.e., antibiotic resistance gene). (**b**) Here we see that the duplication event occurs as homologous regions are recombined mediated by *recA*. This process can result in both increases and decreases in the copy number of the gene of interest. (**c**) Maintaining selection pressure ensures that only cells with the gene of interest and selection marker survive. By subculturing cells in increasing antibiotic concentration, as depicted by the darker shades of green, one can screen for cells with higher gene copy number

desired cassette to a copy number that is proportional to the strength of the selective pressure applied [2]. Strains with a range of copy numbers can then be "fixed" by deletion of *recA*. Because of the ubiquitous nature of *recA*, this process is theoretically applicable in numerous host organisms. Constructed pathways can then be assayed for productivity to identify the most optimal construction, with the best strain carried forward for long-term selection-free fermentation.

2 Materials

Be sure to follow all waste disposal procedures for your institution.

2.1 Construction of Plasmids for Initial Integration

1. Plasmid pTrcHis2B (available from ThermoFisher).
2. Plasmid pAC-LYC (AddGene: Plasmid # 53270) [5] (*see* **Note 1**).
3. *Synechocystis* PCC 6803 (available from ATCC).
4. An instrument for measuring DNA concentration such as a NanoDrop 1000 spectrophotometer.
5. Restriction enzymes: SphI, ClaI, MluI, BamHI, SacI, EcoRI, XbaI.
6. CutSmart® Buffer.
7. Mung bean nuclease.
8. Primers (*see* Table 1).
9. Materials to run a 1% agarose gel, including 1× TAE (Tris–acetate–EDTA) buffer, agarose, 6× DNA gel loading dye, 1 kb DNA ladder, a nucleic acid gel stain such as SYBR Gold or SYBR safe, a gel station, gel imaging system, and a scalpel or razor to excise DNA fragments.
10. Qiagen QIAquick or similar gel extraction kit.
11. Wizard Genomic DNA Purification Kit.
12. PCR Reagent materials, including dNTPs, and high-fidelity polymerase such as PrimeSTAR HS, Q5 or iProof and accompanying buffers.
13. (Optional) Luria broth, agar, plates, ampicillin, and *E. coli* cloning strain DH5α (Invitrogen Corp.).

2.2 Integration of CIChE pTGD Construct into Chromosome

1. *E. coli* strain K12MG1655 (ATCC #700926).
2. Lysogens and transfer strains (From Dr. Dana Boyd and Dr. Jon Beckwith) [6].
3. λ InCh lysates.
4. Phage transduction donor strain: BW26547 *recA::kan* Lambda *recA+* (CGSC *E. coli* Genetic Stock Center, Yale University).
5. *E. coli* DHB6521 (λ InCh strain).

Table 1
Primers for PCR (IDT, Coralville, IA, USA)

Primer	Purpose	Sequence (5′–3′)
Int P1	Amplification of chlB1 (Sense)	GCACCCTACGCATCGCCAGTTCTT
Int P2	Amplification of chlB1 (Antisense)	CGCTCTCGGGCAAACTTTTCTGTGTT
Int P3	Amplification of CAT (Sense)	CGTTGATCGGCACGTAAGAGGTTCC
Int P4	Amplification of CAT(Antisense)	CCTTAAAAAAATTACGCCCCGCCC
Lam P1 (Sense)	Introduction of BamHI to chlB1-CAT (Sense)	TATCGGATCCCCAGTTCTTTCAAAAACGTCCACGCC
Lam P4 (Antisense)	Introduction of SacI/SphI to chlB1-CAT (Antisense)	GGGAGCTCAGCATGCCCTTAAAAAAATTACGCCCCGCCC
Lam P7 (Sense)	Introduction of EcoRI/MluI to chlB copy #2	GGAATTCAACGCGTCCAGTTCTTTCAAAAACGTCCACGCC
Lam P8 (Antisense)	Introduction of XbaI to chlB copy #2	CGTCTAGAGCCCCGTTGTCTTCACTGATCAACACT
Cm qPCR1	Measure genomic CAT copy number (Sense)	CGCCTGATGAATGCTCATCC
Cm qPCR2	Measure genomic CAT copy number (antisense)	AGGTTTTCACCGTAACACGC
BioA qPCR1	qPCR Reference gene to normalize cat copy number to number genomic copy (Sense)	GTGATGCCGAAATGGTTGCC
BioA qPCR2	qPCR Reference gene to normalize cat copy number to number genomic copy (antisense)	GCGGTCAGACGCTGCAACTG

2.3 CIChE Gene Duplication Process

1. Luria–Bertani (LB) broth.
2. 15 mL Falcon culture tubes.
3. Chloramphenicol.
4. Stock of 60% glycerol and cryotube vials.
5. Shaker incubator set at 37 °C.

2.4 recA Deletion Validation by UV Sensitivity

1. BW26547 P1 phage lysate.
2. Stratalinker UV cross-linker model 1800.
3. LB agar plates.
4. Sodium citrate.
5. CrimsonTaq DNA polymerase.

3 Methods

3.1 Construction of Plasmids for Initial Integration (See Note 2)

1. Digest pTrcHis2B with SphI and ClaI, using standard restriction enzyme reaction conditions as dictated by the supplier (1 μg of template DNA, 37 °C, 1 h, using CutSmart® Buffer).
2. Blunt the digested backbone with mung bean nuclease, following manufacturer recommended reaction conditions (0.1 μg/μ L DNA template, 30 °C, 30 min, in CutSmart® Buffer). Deactivate by either phenol–chloroform extraction or by the addition of SDS to 0.01%. Purify by ethanol precipitation.
3. Run a 1% agarose gel to separate digested plasmid fragments and extract the 2.8 kb band containing the pBR322 origin of replication and the β-lactamase (ampicillin resistance) gene (*see* **Note 3**).
4. Self-ligate this blunted 2.8 kb fragment using T4 DNA ligase (*see* **Note 4**)
5. Amplify a ~1 kb region of ChlB from *Synechocystis* PCC 6803 using IntP1 and IntP2 primers (*see* Table 1) (*see* **Note 5**).
6. Amplify chloramphenicol acetyltransferase gene (CAT) from pAC-LYC using primers IntP3 and IntP4 (*see* Table 1) using a high-fidelity DNA polymerase. This will produce a 777 bp fragment (*see* **Note 6**); about any titratable marker will do.
7. Purify products by 1% agarose gel electrophoresis and use a gel purification kit to isolate the DNA fragments.
8. Use overlap PCR to connect the 1 kb region of chlB1 to the CAT.
9. Amplify the constructed chlB1-CAT fragment with primers LamP1 and LamP4 (*see* Table 1) to introduce BamHI and SacI/SphI sites.

10. Cut the truncated pTrcHis2B and the chlB-CAT fragment with BamHI and SacI.
11. Ligate these two fragments together.
12. Amplify a second copy of chlB using LamP7 and LamP8 to introduce EcoRI/MluI and XbaI sites.
13. Cut the resulting chlB fragment and modified pTrcHis2B with EcoRI/XbaI, then ligate together to create pTGD.
14. Insert gene of interest into KpnI/MluI site of this newly created pTGD plasmid (*see* **Note 7**).

3.2 Transformation of ClChE pTGD Construct into Chromosome (See Note 8)

1. Transfer TGD DNA constructs from the TGD platform plasmid to the genome of K12 *E. coli* using the λInCh protocol (Into the Chromosome), which consists of three steps: (*see* **Note 9**).
   (a) Homologous recombination between the pTGD plasmid and a λ phage.
   (b) Integration of the phage-pTGD DNA into the att site of your *E. coli* strain.
   (c) Removal of phage DNA by looping out.

A thorough λInCh protocol can be found at: http://beck2.med.harvard.edu/resources/InCh/download/Lambda_InCh_Manual.pdf.

3.3 ClChE Gene Duplication Process (See Note 10)

1. Grow cells in 5 mL of LB using 14 mL culture tubes overnight at 37 °C with 13.6 μg/mL of chloramphenicol added to the solution until stationary phase.
2. The following day, transfer 50 μL of the cell culture to 5 mL of fresh LB media (1:100) with 27.2 μg/mL chloramphenicol concentration and grow overnight at 37 °C to stationary phase.
3. Repeat **step 12** for several days, each time subculturing the cells into fresh LB with double the concentration of antibiotic until the desired concentration is reached (~**1500** μg/mL) (*see* **Note 11**).
4. At each concentration of chloramphenicol, aliquot a portion of the stationary phase cell culture for *recA* deletion to stabilize the tandem repeats. *RecA* deletion is readily accomplished by P1 phage transduction of the recA::kan allele from BW26547 to the TGD strain using the open access protocol reported from Sauer lab (http://openwetware.org/wiki/Sauer:P1vir_phage_transduction) (*see* **Note 12**; Gene copy stability).

3.4 recA *Deletion Validation by UV Sensitivity*

1. Each population of cells with a *recA*- genotype must be routinely validated using a UV sensitivity assay. To do so, inoculate

10 mL of LB broth with 100 μL of an overnight culture of a *recA*- CIChE strain and grow for 2 h at 37 °C.

2. Measure the concentration of cells by OD_{600}, then centrifuge and wash with 10 mL of M9 minimal medium.
3. Dilute the cells with M9 to a density of ~100 colonies per sample and place on two LB agar plates.
4. Expose one of the two plates to UV using a device, such as a Stratalinker UV cross-linker model 1800, in a darkened room at 3000 μJ energy, and set the second aside to serve as a control.
5. Incubate both plates overnight at 37 °C in the dark and the following day, determine the number of colonies on each plate to calculate the corresponding cell survival. *recA*- strains are very sensitive to UV exposure, and UV exposed *recA*- strains should have substantially reduced colonies compared to WT.
6. *recA*- strains can then be grown and cryostored for further use.
7. Verify copy number quantification can be accomplished using qRT-PCR on the amplified genomic DNA TGD cassette relative to a reference gene by growing post CIChE *recA*- variants in LB overnight at 37 °C, then use the Promega Wizard Genomic DNA purification kit to isolate genomic DNA containing the amplified cassette (*see* **Note 13**).
8. Use a standard qRT-PCR SYBR green method to quantify the CAT gene in the amplified cassette DNA relative to the *E. coli* genomic DNA gene *BioA*, with gene specific primers (*see* Table 1).

4 Notes

1. This plasmid is used here for its chloramphenicol resistance marker. Other sources of the CAT gene are sufficient substitutes, although this would likely alter the primers necessary for the protocol.
2. There are multiple options for constructing the cassette. Theoretically, much of the tandem gene duplication cassette could be directly synthesized or assembled with other methods than the one outlined. However, due to the homologous flanking regions, some level of restriction digestion and ligation is necessary. In addition, it should be ensured there are no other tandem repeats in the cassette that may result in duplication of the antibiotic resistance marker without duplicating the entire desired region inside the two homologous repeats. This is critical to the experiment's success.
3. Anecdotally, we have observed that freezing excised gel fragments overnight at −20 °C has produced better gel extraction

yields. If the gel extraction product is giving poor OD 230/260 readings, consider allowing wash buffers to site on the column 2–5 min before spin down. In addition, consider adding a second loading buffer wash in addition to the ethanol washes.

4. You may have a low yield of this ligated product or simply want more of this product to use for different steps downstream. If so, at this step you could transform this product and plate on ampicillin, pick colonies, grow them in LB with ampicillin, and perform a plasmid miniprep to get more of this product.
5. The regions of homology dictated here have been validated for use in *E. coli*. Certainly, other ~1 kb segments could be sufficient. We provide this segment of DNA as an example. In addition, in lieu of obtaining this entire strain, just the ChlB gene could be directly synthesized and carried forward for subsequent steps.
6. As mentioned, other sources of CAT are possible. In fact, other selective markers are also usable as long as their effect is titratable [7].
7. Remember to include any necessary regulatory elements along with your gene of interest in your cassette design, including a promoter.
8. Be certain the strain you are transforming into is recA+, otherwise recombination will not occur. Also, be sure your background strain can tolerate high expression of your gene.
9. Other chromosomal integration techniques are possible, including that described by Datsenko and Wanner [8].
10. Once this CIChE process has begun, it will take several days of maintenance.
11. If cells are unable to grow to stationary phase overnight, allow to continue growing for an additional 24 h and/or subculture 50 μL into a fresh 5 mL of the same concentration of chloramphenicol supplemented LB. At higher antibiotic concentrations, it is recommended to subculture the evolving population into both double and 1.5×, antibiotic concentration with respect to the previous culture.
12. Gene copies are not stable during amplification. A mutation in one copy can propagate quickly causing many copies to be inactive. Make freezer stocks of strains after each subculture as backups, and only do the minimum number of cell doublings required to achieve the copy number you desire.
13. Select several colonies from each population for qPCR characterization, as not all variants within a given antibiotic concentration culture will have the same copy number.

Acknowledgments

National Science Foundation CAREER (1452549), National Institutes of Health (NIH) Common Fund (5R01MH103910-02) and Biotechnology Training Grant (5T32GM008449-23).

References

1. Biggs BW, De Paepe B, Santos CNS, De Mey M, Kumaran Ajikumar P (2014) Multivariate modular metabolic engineering for pathway and strain optimization. Curr Opin Biotechnol 29:156–162
2. Tyo KEJ, Ajikumar PK, Stephanopoulos G (2009) Stabilized gene duplication enables long-term selection-free heterologous pathway expression. Nat Biotechnol 27:760–765
3. Kowalczykowski SC, Dixon DANA, Eggleston AK, Lauder SD, Rehrauer WM (1994) Biochemistry of homologous recombination in Escherichia coli. Microbiol Rev 58:401–465
4. Zhang J (2003) Evolution by gene duplication: an update. Trends Ecol Evol 18:292–298
5. Cunningham FXJ, Sun Z, Chamovitz D, Hirschberg J, Gantt E (1994) Molecular structure and enzymatic function of lycopene Cyclase from the Cyanobacterium Synechococcus sp strain PCC7942. Plant Cell 6:1107–1121
6. Boyd D, Weiss DDS, Chen JJC, Beckwith J (2000) Towards single-copy gene expression systems making gene cloning physiologically relevant: lambda InCh, a simple Escherichia coli plasmid-chromosome shuttle. J Bacteriol 182:842–847
7. Chen Y-Y, Shen H-J, Cui Y-Y, Chen S-G, Weng Z-M, Zhao M, Liu J-Z (2013) Chromosomal evolution of Escherichia coli for the efficient production of lycopene. BMC Biotechnol 13:6
8. Datsenko KA, Wanner BL (2000) One-step inactivation of chromosomal genes in Escherichia coli K-12 using PCR products. Proc Natl Acad Sci 97:6640–6645

Chapter 5

Bacterial Expression of Membrane-Associated Cytochrome P450s and Their Activity Assay in Nanodiscs

Demetrios Maroutsos, Hannah Huff, and Aditi Das

Abstract

Eukaryotic membrane bound cytochrome P450s are expressed in bacterial systems to produce large yields of catalytically active protein for structure function studies. Recently, there have been several instances of expressing eukaryotic membrane bound CYPs in bacteria after making various modifications to both the N-terminus membrane binding domains of the protein and to noncontiguous F-G membrane binding loop that is also implicated in substrate binding. These modifications have been shown not to disturb the function of the protein of interest. The major factors that have been key to express the membrane bound cytochrome P450s in bacteria have been the following: (a) exon optimization (b) selection of the appropriate vector and host strain, and (c) growth and expression conditions with respect to temperature and speed of shaking the media flask. Herein, we describe methods to express and purify eukaryotic membrane bound cytochrome P450s. We also describe the measurement of the activity of the cytochrome P450 expressed by taking the example of cytochrome P450 2J2, the primary P450 expressed in the human heart and CYP725A4, the primary cytochrome P450 expressed in the first step of taxol synthesis. Additionally, we discuss the pros and cons of the different modifications done in order to express the membrane bound cytochrome P450s.

Key words Cytochrome P450, Protein purification, Activity assay, CYPs, His-Tags, N-terminal modification

1 Introduction

Cytochrome P450s (CYPs) are a class of membrane-bound enzymes responsible for the synthesis of numerous steroid hormones, fatty acids, eicosanoids and bile acids. They are also involved in the metabolism of drugs and xenobiotics in the body. Hence, CYPs play diverse chemical roles within the cell as they can metabolize a large variety of substrates. In order to metabolize substrates, CYPs interact with redox partners such as cytochrome P450 reductase (CPR) that transfers electrons from NADPH to CYPs to facilitate substrate metabolism. While most of the prokaryotic CYPs are soluble, all the eukaryotic CYPs are membrane bound

Christine Nicole S. Santos and Parayil Kumaran Ajikumar (eds.), *Microbial Metabolic Engineering: Methods and Protocols*, Methods in Molecular Biology, vol. 1927, https://doi.org/10.1007/978-1-4939-9142-6_5,

and display higher complexity and chemical versatility as compared to soluble P450s.

The activity of these membrane bound CYPs can be modulated by the lipid composition of the membranes. As the eukaryotic CYPs are membrane bound, their overexpression in bacterial systems required several modifications including N-terminus truncations, hydrophilic residue replacement in the N-terminus, silent mutations to change the codon usage are necessary adjustments in order to optimize protein expression, to conserve proper folding and to maintain its natural bioactive function when expressed in bacterial systems such as *E. coli*. The use of recombinant bacterial systems for expression of CYPs provides an avenue for inexpensive large scale production of proteins for downstream biochemical studies including crystallization [1].

Briefly, this chapter describes how a particular eukaryotic CYP gene is ligated into a bacterial expression plasmid of choice and transformed into competent *E. coli* cells. The lac promoter precedes the P450 gene as it can be easily induced by the nonhydrolyzable analog to lactose, namely, isopropyl ß-d-1-thiogalactopyranoside (IPTG). By inducing the cells at an optimal log phase of bacterial growth, the expression yields of functioning membrane protein is increased.

In order to facilitate the purification of the target CYP from bacterial expression systems, a polyhistidine tag (His-tag) is biochemically attached to either the N or C terminus of a protein. Moreover, silent mutations typically in the N-terminus are necessary to relieve expression difficulties within the bacterial system. Codons for the same amino acid sequences are used in different frequencies within different species. Silent mutations help relieve pauses in translation in the bacterial cell and ease the process altogether. The pauses are due to a lower concentration of tRNA with the anticodon to the mRNA copy, and they can foster improper folding of the protein of interest. Silent mutations do not alter the protein structure as they are the same amino acid, but this methodology relieves the translational difficulties [1].

The N-terminal domain for eukaryotic cytochrome P450s is important in that it can mediate association to the lipid membrane, it can serve as the entrance for substrates, and it can be used for binding to other proteins for downstream reactions [2, 3]. After removal of the N-terminus, CYP2C5 and CYP2C3 were shown to become solubilized, as the lipid-associating part of the protein had been cleaved [4]. However, it is not always only necessary to truncate the N-terminus to achieve solubility, as there are other hydrophobic residues that affect membrane association. It has also been shown that N-terminal cleavage disrupts CYP-CYP interactions—as seen in CYP3A4 and CYP2C9's ability to dimerize [5]. Also, modifications to the N-terminus of CYP1A2 were shown to affect the catalytic activity of the enzyme as well as the

CYP's affinity to the composition of lipids in the membrane. In CYP2J2, the N-terminus hydrophobic residues are used for attaching to the lipid membrane. In a mutational study aiming to optimize the expression of CYP2J2 in *E. coli,* different constructs were prepared. In one of the constructs, the second codon was replaced with an alanine, silent mutations were introduced in the first 5–7 codons to make them AT rich and ease mRNA secondary structures, a truncation of most of the membrane spanning region and the incorporation of the sequence AKKTSS (hydrophilic) on the N-terminus. Importantly, the incorporation of the sequence "LLLAVFL" is a ribosome binding consensus sequence and that helps the mRNA get translated a lot more efficiently across many CYP types. In CYP725A4, a CYP that metabolizes taxol, used N-terminus mutations to achieve maximum functional expression in *E. coli* cells, uses the similar mutations as did CYP2J2. All in all, the modifications of the N-terminus are important in maximizing the expression for producing a high yield. This chapter focuses on the expression, purification, and activity measurement of two specific eukaryotic cytochrome P450s—human CYP2J2 and taxol plant CYP725A4 [6] in *E. coli.*

2 Materials

2.1 Materials for CYP2J2 Expression, Purification, and Activity Analysis

2.1.1 Expression and Purification of CYP2J2

1. Cell Strain: DH5α.
2. Vector: pCWori.
3. CYP2J2 cDNA and pTGro7 gene (Origene).
4. Luria–Bertani broth (LB): 10 g tryptone, 5 g yeast extract, 10 g NaCl. Autoclave on liquid setting for 20 min, no dry time.
5. Ampicillin: 1 g of ampicillin in 10 mL of H_2O (0.1 g/mL), filter-sterilize. Store at −20 °C.
6. Chloramphenicol: 0.2 g of chloramphenicol in 10 mL of ethanol (20 mg/L), filter-sterilize. Store at −20 °C.
7. Terrific Broth: 12 g tryptone, 12 g yeast extract, 2 g peptone, and 4 mL of glycerol in 1 L of water, autoclaved.
8. KP_i Buffer: 1 M K_2HPO_4, 1 M KH_2PO_4; titrate to get pH 7.4. Autoclave and store at 4 °C.
9. Delta-aminolevulinic acid.
10. Cary 300 UV-Vis spectrophotometer.
11. IPTG: (isopropyl β-D-1-thiogalactopyranoside), 1 M stock in MilliQ water, filter-sterilize, store at −20 °C.
12. Arabinose.
13. Trace Minerals Spike: Recipe (per liter)—400×: 27.0 g $FeCl_3 \cdot 6H_2O$–2.0 g $ZnCl_2 \cdot 4H_2O$–2.0 g $CoCl_2 \cdot 6H_2O$–2.0 g

Na_2MoO_4–1.0 g $CaCl_2$. $2H_2O$–1.0 g $CuCl_2$–0.5 g H_3BO_3–100 mL Conc. HCl.

14. Sorvall RC 6+ centrifuge.
15. Quartz cuvettes.
16. Buffer A (recipe described in protocol).
17. Beckman ultracentrifuge Ti-45 rotor.
18. Magnetic Stirrer and small cylindrical stir bar.
19. Buffer B (recipe described in protocol).
20. EDTA.
21. Ni-NTA 5 mL column (Gold Biotechnology).
22. Column Buffer C (recipe described in protocol).
23. Cholate detergent.
24. ATP.
25. Amicon centrifugal filter (10,000 or 30,000 MWCO).
26. Imidazole.
27. Millipore H_2O.

2.1.2 Spectrophotometric Characterization and Measurement of Mutant Activity

1. CARY 300 UV-Vis spectrophotometer (Agilent).
2. Suba seal septa (Sigma).
3. Parafilm.
4. 0.1 M potassium phosphate (KPi) buffer, prepared as mentioned previously.
5. Ultrapure N_2 gas.
6. Anaerobic chamber (Coy).
7. Glass-coated stir bar.
8. Carbon monoxide gas.
9. Safranin T.
10. Millipore H_2O.
11. Purified CYP2J2 of known concentration.
12. Nail polish—any brand.
13. Dithionite.
14. Quartz cuvettes—2.
15. 100 μL Hamilton syringe.
16. Cary WinUV software.
17. NADPH.
18. Ebastine (Santa Cruz Biotechnology).
19. Purified CPR of known concentration.
20. Ammonium acetate.

21. Glacial acetic acid.
22. Methanol.
23. Acetonitrile.
24. Waters 2695 Separations Module.
25. Analytical cyano column (Alltech; 250 mM × 4.6 mM; Cat. #: 605CN).
26. 0.45 μm centrifugal filters (VWR).

2.1.3 Incorporation of CYP2J2 into Nanodiscs

1. 1-palmitoyl-2-oleoyl-*sn*-glycero-3-phosphocholine (POPC; Avanti Polar Lipids).
2. 1-palmitoyl-2-oleoyl-*sn*-glycero-3-phospho-L-serine (POPS; Avanti Polar Lipids).
3. Any other desired lipids (e.g., sphingomyelin, cholesterol).
4. Cholate.
5. Millipore H_2O.
6. Purified MSP of known concentration (*see* previous sections).
7. Purified CYP2J2 of known concentration (*see* previous sections, does not have to be buffer exchanged).
8. Bio-Beads (Bio-Rad or Amberlite XAD-12 (Sigma).
9. Glycerol.
10. Potassium phosphate buffer, prepared as mentioned previously.
11. 0.45 μm centrifugal filters (VWR).

2.1.4 Expression and Purification of Cytochrome p450 Reductase (CPR)

1. Cell strain: DH5α.
2. Plasmid: pOR262 from Todd Porter; expressed protein contains ompA signal peptide.
3. Ampicillin.
4. LB media, prepared as previously mentioned with adjustments for volume.
5. Riboflavin (1 mg/mL) solubilized in dd H_2O.
6. Isopropyl β-D-1-thiogalactopyranoside (IPTG), stored as a 1 M stock.
7. Tris–HCl.
8. Sucrose.
9. Ethylenediaminetetraacetic acid (EDTA).
10. Lysozyme.
11. DNase I.
12. Phenylmethanesulfonyl fluoride (PMSF).
13. Triton.
14. Glycerol.

15. Dithiothreitol (DTT).
16. Triton.
17. $NADP^+$.
18. Potassium ferricyanide—$K_3[Fe(CN)_6]$.
19. Cytochrome c (Sigma).
20. Purified CPR of known concentration.
21. NADPH.
22. Potassium phosphate buffer, prepared as mentioned previously.
23. Millipore H_2O.
24. Cary 300 UV-Vis spectrophotometer (Agilent).

2.2 Materials for CYP725A4 Expression and Purification

2.2.1 Expression and Purification of CYP725A4

1. Strain: *E. coli* strain K12 MG1655.
2. LB Broth (recipe mentioned above).
3. Spectinomycin.
4. Ampicillin.
5. Terrific Broth (recipe mentioned above).
6. Delta-aminolevulinic acid.
7. Cary 300 UV-Vis spectrophotometer.
8. IPTG: (isopropyl β-D-1-thiogalactopyranoside), 1 M stock in MilliQ water, filter-sterilize, store at −20 °C.
9. Arabinose.
10. Shaker incubator.
11. Sorvall GSA rotor (Thermo Scientific).
12. Lysozyme buffer (recipe mentioned in protocol).
13. Sonicator (a Heat Systems Ultrasonics (Qsonica, Newtown, CT).
14. Beckman Ti-45 rotor (Beckman, Coulter, Brea CA).
15. Extraction buffer (recipe mentioned in protocol).
16. Ni-NTA column (GoldBio, St. Louis, MO).
17. KP_i buffer.
18. EDTA.
19. DTT.
20. Imidazole.
21. Amicon Ultra 10,000 MWCO concentrator.
22. Amicon Ultra 50,000 MWCO concentrator.

2.2.2 Expression and Purification of Stevia CPR

1. LB media, prepared as mentioned previously.
2. TB media, prepared as mentioned previously.
3. Ampicillin.

4. Riboflavin.
5. Isopropyl β-D-1-thiogalactopyranoside (IPTG).
6. Lysozyme buffer, prepared as mentioned previously for CPR (Note: any deviations from any standard buffer recipe will be mentioned in the Subheading 2).
7. Lysozyme.
8. Lysis buffer, prepared as mentioned previously.
9. Column buffer, prepared as mentioned previously.
10. ADP-agarose column.
11. Sodium chloride.
12. Dithiothreitol (DTT).
13. Cholate.
14. Tergitol.
15. Riboflavin-5′-phosphate.
16. Adenosine.
17. NADPH.
18. Cary 300 UV-Vis spectrophotometer.
19. Purified CYP725A4 of known concentrations.
20. Purified RnCPR and SnCPR of known concentrations.
21. 0.25 M Tris–HCl buffer, pH 7.5.
22. Flavin adenine dinucleotide (FAD).
23. Flavin mononucleotide (FMN).
24. Glucose-6-phosphate.
25. Glucose-6-phosphate dehydrogenase.
26. Taxadiene.
27. Hexanes.
28. Ether.
29. Purified MSP1E3D1 of known concentration.
30. Amberlite XAD-12 beads (Sigma).
31. Potassium phosphate (KPi) buffer, prepared as mentioned previously.
32. Ni-NTA agarose column, GoldBio.

3 Methods

3.1 Methods for CYP2J2 Expression, Purification, and Activity Analysis

3.1.1 Expression and Purification of His Tagged CYP2J2

Construction of CYP2J2 ***N-deletion mutants and heterologous expression in E. coli*****:** The native human CYP2J2 gene was obtained from OriGene Technologies. The Kyte-Doolittle hydropathy plot was used to find out that approximately the first 33 amino acids of CYP2J2 form the hydrophobic membrane-spanning domain. These membrane spanning domains vary from protein to protein. Based on literature and rationale that is described in our review paper [1], we prepared five different CYP2J2 constructs with varying N-terminal modifications for the purpose of increasing the yield of expression without eliminating their ability to bind native substrate or associate with the lipid bilayers. The different N-terminally modified CYP2J2 constructs are listed in a previous paper [2]. The following N-terminal modifications were made to the CYP2J2 gene construct. (1) The second codon in the gene sequence was replaced with Ala (GCT), as many E. coli genes controlled by the lacZ operator have been found to contain this codon. (2) Several silent mutations were incorporated in the first five to seven codons to render them both AT rich and to minimize the free energy of mRNA secondary structure. This also matches with *E. coli* codon usage. (3) The hydrophobic N-terminus was truncated and best yields were obtained when the N-terminus was fully truncated. Finally, (4) hydrophilic sequence AKKTSS was incorporated into the N-terminus. Previously, the incorporation of these residues has aided in the expression of several other cytochrome P450s. One of the complications in the molecular cloning of CYP2J2 was the presence of an internal NdeI site in the gene that was digested with a different enzyme, AseI, which produced an overhang compatible with the NdeI digested fragment of the vector and enabled ligation producing the desired constructs without the NdeI site. The gene constructs of CYP2J2 were transformed into DH5alpha cells along with the pTGro7 plasmid (encoding for the expression of the GroEL-GroES chaperonin system). The maximum yield was reported for M2D34G where the complete truncation of the N-terminal hydrophobic region of CYP2J2 was accomplished. The oligomerization studies showed that the complete deletion of the N-terminus leads to production of the monomeric state of the protein.

3.1.2 Expression of CYP2J2 in 6 Flasks with 500 mL Media in each Flask

1. Cotransform DH5α cells with cDNA of modified CYP2J2 on a pCWori vector along with the pTGro7 gene using standard protocols for expression of the enzyme of interest as well as the chaperonin system that will facilitate proper folding of CYP2J2 [2].
2. After recovery, streak out double transformants on an LB plate with chloramphenicol (100 μg/mL) and ampicillin (20 μg/mL).

The streak should aim to produce individual colonies, thus, streak the plate out using a sterile inoculating loop and moving the cells back and forth whilst rotating the plate (*see* **Note 1**).

3. Incubate the plate with the agar side facing upward (cells down) at 37 °C for 12 h.
4. Create multiple 25 mL solutions of Luria broth (LB) media in 125 mL Erlenmeyer flasks (covered with foil) and autoclave. Once cool, inoculate them with 100 μg/mL of chloramphenicol and 20 μg/mL of ampicillin (*see* **Note 2**).
5. Pick up one colony from the plate using a sterile, disposable tip of a micropipette. Drop the whole pipette tip into the 25 mL LB media and re-cover flask. Do this for 2–3 flasks, all under flame (*see* **Note 3**).
6. Incubate the Erlenmeyer flasks with the cells at 37 °C at 220 rpm overnight (15–16 h). This is termed as the starter culture.
7. For 6 flasks, prepare 2700 mL of Terrific Broth (TB) media and 300 mL of 1 M potassium phosphate buffer (KP_i) at pH 7.4. Place 450 mL of the TB each into six separate 2.8 L Fernbach flasks and cover the mouth of the flask with aluminum foil, autoclave the KP_i separately, and also autoclave a graduated cylinder with aluminum foil that will be used to add the KP_i to the larger media after autoclaving.
TB Recipe for 500 mL culture volume

6 g tryptone.

6 g yeast extract.

1 g peptone.

2 mL of glycerol.

448 mL of MilliQ deionized water per 450 mL.

Recipe for Trace Minerals Spike

Trace Minerals Spike Recipe (per Liter)—400×.

27.0 g $FeCl_3$. $6H_2O$–2.0 g $ZnCl_2.4H_2O$–2.0 g $CoCl_2.6H_2O$–2.0 g Na_2MoO_4–1.0 g $CaCl_2$. $2H_2O$–1.0 g $CuCl_2$–0.5 g H_3BO_3–100 mL Conc. HCl.

8. Add the following to each 2.8 L Fernbach flask under a flame, flaming the lip of the flask upon addition of each component.

450 mL of TB (already in flask).

50 mL of KP_i pH 7.4.

0.5 mL of 1000× Ampicillin.

0.5 mL of 1000× Chloramphenicol.

0.125 mL of 4000× trace minerals

9. Add 5 mL (1/100 volume) of overnight starter culture (i.e., 5 mL into 500 mL media), under a flame into each of the 2.8 L flasks (*see* **Note 4**).
10. Grow the 500 mL cultures for 2.5 h at 37 °C and 220 rpm.
11. After 2.5 h, add 500 μL of delta-aminolevulinic acid (δ-ALA) to each flask (from a 0.5 mM stock) and grow for 2 h at 26 °C at 160 rpm (*see* **Notes 5** and **6**).
12. Grow the cells and take a sample from one of the flasks regularly until the absorbance at 600 nm is 1.2. Upon reaching an OD_{600} of 1.2, induce each flask with 1 mM IPTG (isopropyl β-D-1-thiogalactopyranoside) and 2 g of Arabinose added under flame (*see* **Note 7**).
13. Grow the cultures for 44 h at 26 °C and 160 rpm.
14. Harvest the cells by centrifugation at 9600 × *g* for 15 min at 4 °C; discard the supernatant and scrape out the pellet into a freezer-safe plastic bag (*see* **Note 8**).
15. Freeze the cell pellet at −80 °C and store it in freezer bag or proceed with purification as outlined below.

3.1.3 Purification of CYP2J2

1. Resuspend the cell pellet in a beaker with buffer A that is 10% of the culture volume (e.g., 100 mL of buffer for cell pellet from 1 L culture) (*see* **Note 9**).
   Recipe of Buffer A (300 mL for 3 L culture volume, adjust if other culture volume used).

   75 mL 80% Glycerol (final 20% glycerol).

   30 mL 1 M KPi (final 0.1 mM KPi).

   900 μL 2 M $MgCl_2$ (final 6 mM $MgCl_2$).

   105 μL βME (final 5 mM βME).

   600 μL 0.1 M PMSF (final 0.2 mM PMSF).

   194 mL MilliQ H_2O.

   ~1 mg each of DNase and RNase.

2. Sonicate on ice for six cycles of 40 s, with breaks of 40 s at an output of 4, 40% duty cycle (a Heat Systems Ultrasonics; Qsonica, Newtown, CT) (*see* **Note 10**).
3. Centrifuge the sonicated cells at 142,000 × *g* at 4 °C for 45 min using a Beckman ultracentrifuge Ti-45 rotor, followed by collecting the membrane pellet.
4. Resuspend the membrane pellet in solubilization buffer B for 3–5 h with stirring at 4 °C with 10% of the culture volume.

Recipe of Buffer B (300 mL for 3 L culture volume, adjust if other culture volume used).

75 mL 80% Glycerol (Final 20% Glycerol).

30 mL 1 M KPi (final 0.1 M Kpi).

15 mL 4 M NaCl (final 0.2 M NaCl)

105 μL βME (final 5 mM βME)

30 mL 10% Cholate (final 1% Cholate)–add after pellet has resuspended in the buffer.

150 mL MilliQ H_2O

5. Centrifuge the suspended membrane pellet at 142,000 × *g* at 4 °C for 45 min using a Beckman ultracentrifuge Ti-45 rotor followed by collecting the supernatant that will be further purified using a Ni-NTA column (*see* **Note 11**).
6. Recharge the 25 mL of Ni-NTA column by first adding 2 mL of 0.5 M EDTA, and washing with 25 mL of MilliQ H_2O followed by charging the Ni-NTA column with 2 mL of 0.1 M $NiSO_4$. Wash with 25 mL of MilliQ H_2O until there is no more colored flow-through from the column.
7. Equilibrate the column with column buffer C:
   Recipe of Column Buffer C (100 mL).

25 mL 80% Glycerol (final 20% glycerol).

10 mL KPi (final 0.1 M KPi).

1 mL 10% Cholate (final 0.1% Cholate).

64 mL MilliQ H_2O.

3.5 μL βME (final 0.5 mM β-ME)–add at the time of purification.

8. Load the supernatant from **step 5** onto the charged Ni-NTA column (*see* **Notes 12** and **13**).
9. Wash the column with five column volumes of column buffer.
10. Wash with four column volumes of ATP-containing wash buffer prepared as follows:
    (a) 100 mM potassium phosphate pH 7.4.
    (b) 20% glycerol.
    (c) 0.1% cholate.
    (d) 5 mM ATP.
    (e) 10 mM magnesium chloride.
    (f) 150 mM KCl.

This buffer removes the GroEL that may be bound to the CYP proteins. This might require incubating for more membrane-bound CYP proteins.

11. Wash the column with one column volume of Column Buffer that includes 20 mM imidazole.
12. Elute the protein with four volumes of the elution buffer prepared as follows:
    (a) 100 mM potassium phosphate pH 7.4.
    (b) 20% glycerol.
    (c) 0.1% cholate.
    (d) 200 mM imidazole.
    (e) 200 mM NaCl.
13. Collect multiple fractions of 2–4 mL each until the fractions are colorless. CYP2J2 containing fractions will have an obvious red tint.
14. Concentrate the protein down using an Amicon centrifugal filter (10,000 or 30,000 MWCO). A buffer exchange can be performed to remove the imidazole (*see* **Note 14**).

3.1.4 Spectrophotometric Characterization of CYP2J2 N-terminus Mutants and Measurement of Their Activity

CYP2J2 concentration was measured by carbon monoxide difference spectra using an extinction coefficient of 91 mM^{-1} cm^{-1}. CYP2J2 is similar to other P450s which have typical peaks at 417 nm (Soret band) and the α and β bands (Q-bands) at 536 and 570 nm, respectively. CYP2J2 is reduced with dithionite in the presence of carbon monoxide (CO) and the Fe(II) CO spectra is measured for each of the constructs. The folded CYP2J2 shows a peak at 450 nm.

Binding of substrates to cytochrome P450 is typically accompanied by a spin state shift that leads to the shift in the Soret band from 417 to 393 nm. CYP2J2 is involved in the metabolism of certain drugs such as ebastine. The binding of ebastine to CYP2J2 is calculated by monitoring the spin state shift using a spectrophotometer. Samples contained 2–5 μ*M* of CYP2J2, 100 mM phosphate buffer (pH 7.4), and ebastine. Ebastine is added incrementally from a 10 mM ethanol stock with the final concentration of ethanol not exceeding 1%. The absorption spectra is measured after each ebastine titration. For each titration point, the characteristic type I spectral shift difference, from a low spin to a high spin state, is calculated (by measuring difference $A_{386} - A_{417}$) and plotted against the ebastine concentration that is used. The data is fitted to a standard binding equation.

In order to measure the activity of CYP2J2, the rate of NADPH oxidation and ebastine hydroxylation or arachidonic acid metabolism is measured. Furthermore, the coupling efficiency is measured, which is defined as the ratio of rate of product formation to the rate of NADPH oxidation. The CYP2J2-CPR system can be a detergent solubilized lipid reconstituted system or CYP2J2-CPR incorporated into nanoscale lipid bilayers known as nanodiscs. CYP2J2 constructs in Nanodiscs typically exhibit similar catalytic rates to what is reported for microsomal preparations of CYP2J2 from insect and mammalian expression systems.

3.1.5 Spectroscopic Characterization of CYP2J2 N-terminus Mutants Using CO Binding Assay

1. Turn on the UV-Vis spectrophotometer (CARY 300) and allow the lamp to warm for at least 15 min before use in order to obtain stable spectra.
2. Take two 50 mL aliquots of 0.1 M KPi buffer in 125 mL conical flasks and seal both with Suba seal septa and Parafilm on the top.
3. Prepare anaerobic buffer by purging this 50 mL KPi sealed buffer with inert gas (argon or ultrapure nitrogen) for 20 min while stirring with a glass stir bar. Seal septa with nail polish immediately after removal of purging needles to prevent introduction of oxygen. Place buffer in COY anaerobic chamber when finished (*see* **Note 15**).
4. Prepare CO saturated buffer by purging 50 mL of KPi sealed buffer with inert gas for 20 min, followed by purging with carbon monoxide for 30 min. Seal septa with nail polish immediately after removal of purging needles to prevent introduction of oxygen.
5. Dissolve 0.4 mg of Safranin T in 1 mL of H_2O and dilute 100-fold to make a working solution.
6. Measure the concentration of CYP2J2 and combine 750 pmol protein with KPi buffer to a final volume of 300 μL. Add 6.34 μL Safranin T working solution as a redox mediator to facilitate reduction of CYP2J2.
7. Purge protein solution with inert gas for 20 min before transferring to the COY anaerobic chamber, along with all necessary tools (*see* **Note 16**).
8. In the anaerobic chamber, prepare an anaerobic solution of dithionite in 3 mL anaerobic KPi buffer (*see* **Note 17**).
9. Combine 200 μL purged CYP2J2 solution with 200 μL anaerobic KPi buffer and 20 μL dithionite solution in a quartz cuvette. Seal the cuvette with suba seal, mix well, and remove from the anaerobic chamber.
10. Blank the spectrophotometer using KPi buffer (aerobic).
11. Read the absorption spectra of the dithionite reduced CYP2J2.
12. Add 100 μL of CO-saturated KPi buffer to this using a gas-tight Hamilton syringe to pierce the septa. Read the absorption spectra. Immediately seal the septa of the CO-saturated KPi buffer with nail polish if it is to be used for multiple assays.
13. Compute difference spectra by subtracting absorbance of the reduced protein from the absorbance of the CO-bound protein. The difference spectra should be recorded to quantify the protein as well as to assess the quality of the protein.

3.1.6 Measurement of Activity and Metabolism of CYP2J2

1. Turn on a UV-Vis spectrophotometer at least 15 min before use. Set the spectrophotometer to maintain a temperature of 37 °C.
2. Blank the spectrophotometer using 0.1 M KPi buffer.
3. In a clean quartz cuvette, combine 50 pmol CYP2J2, 150 pmol CPR (*see* next section for the purification of CPR the redox partner), 25 μL 1 M KPi buffer, 10 nmol ebastine, and MilliQ H_2O for a total volume of 245 μL.
   (a) After H_2O and KPi, add the components in the following order: CYP2J2, CPR, ebastine.
4. After adding ebastine, place the cuvette in the spectrophotometer to incubate at 37 °C for 5 min.
5. After 5 min, add 5 μL of 10 mM NADPH and mix quickly with a Pasteur pipette (*see* **Note 18**).
6. Immediately after mixing, monitor the decay of NADPH absorption at 340 nm for 10 min.
7. At the 10-min, quench the reaction with 250 μL ethyl acetate or acetonitrile, depending on the next step of analysis.
   (a) If metabolites are to be analyzed using HPLC, quench with ethyl acetate.
   (b) If metabolites are to be analyzed using LC-MS, quench with acetonitrile.
8. Transfer the quenched sample from the CYP2J2 activity assay to a clean glass tube and add 100 μL each of 4 M NaCl and glacial acetic acid. Vortex for 1 min to mix.
9. Centrifuge for 5 min at 1150 × *g* and 4 °C.
10. Transfer the organic (top) layer to a clean glass tube and set aside (*see* **Note 19**).
11. Add 250 μL ethyl acetate to the aqueous layer, vortex, and centrifuge again.
12. Again transfer the organic layer to a collection tube.
13. Repeat **steps 11** and **12** once more, accumulating the organic layer in one tube.
14. Dry the organic layer on a rotary evaporator or under a stream of inert gas. At this point, samples may be stored at −20 °C if they are not to be analyzed shortly. Otherwise, they may be resolubilized and analyzed as follows.
15. Prepare acetate buffer by combining 0.5 g ammonium acetate, 5 mL glacial acetic acid, and 520 mL dd H_2O.
16. Prepare mobile phase by combining acetate buffer with methanol and acetonitrile in a 48:30:20 (Acetate–MeOH–ACN, v/v/v) ratio.

17. Resolubilize the organic layer in 100 μL mobile phase by quickly vortexing. Filter the solubilized organic layer through a 0.45 μm centrifugal filter tube. Place tube on ice until needed.
18. Set up an HPLC method using the following parameters:
    (a) Time: 30 min.
    (b) Gradient: linear.
    - Start: 100% A (mobile phase) /0% B (acetonitrile).
    - End: 50% A/50% B.

    (c) Temperature: 40 °C for column and sample.
    (d) Column.
    - Column Size: 250 mM × 4.6 mM.
    - Type: Cyano.
    - Particle Size: 5 μm.
19. Inject sample onto HPLC to quantify production of metabolites and remaining substrate at 254 nm.
    (a) A standard curve should be created with known amounts of ebastine before injecting unknown samples.

3.1.7 Incorporation of CYP2J2 into Nanodiscs

Nanodisc technology has been developed to provide a native-like membrane environment in which membrane-bound proteins such P450s can be stable and biophysical techniques can be applied to the protein. Nanodiscs consists of membrane scaffold protein (MSP) and various phospholipids, which are self-assembled into soluble disc-shaped bilayers upon the removal of lipid-solubilizing detergent. The length of the surrounding MSP determines the size of the nanodisc, and can consequently affect the oligomerization state of any incorporated P450s. Use of MSP1D1 will produce 10 nm nanodiscs, while MSP1E3D1 will produce 13 nm nanodiscs. During the assembly mixture, the concentration of the glycerol has to be below 3% and cholate concentration has to be minimum 20 mM. After the self-assembly process as detailed in the methods section, nanodiscs are purified by fractionation on a size-exclusion column. This produces two peaks representing protein-containing nanodiscs and empty discs, respectively, which can be easily differentiated by the characteristic P450 absorbance at 417 nm, however peak differentiation will only occur if the protein is of considerable size. The fractions containing the CYP2J2 nanodiscs are collected and concentrated and used in CYP2J2 assay.

3.1.8 Incorporation of CYP2J2 into Nanodiscs

1. Aliquot lipids into a clean glass tube according to desired nanodisc composition and dry on a rotary evaporator. Place tube in a vacuum desiccator and dry at least 4 h, preferably overnight (*see* **Note 20**).

2. To resolubilize lipids, add 100 mM cholate to make the total lipid concentration 50 mM, vortex the tube, and run briefly under hot water (optional). To ensure complete solubilization, sonicate the tube for 10 min in a water bath.
3. Add MSP solution to yield the desired lipid to scaffold protein ratio. Incubate for 15 min while gently rocking (ratio and temperature varies by main lipid component). If the discs are to be empty, skip **step 4** and proceed to **step 5** (*see* **Note 21**).
   (a) MSP1E3D1 ratios are double those of MSP1D1.

Lipid	Ratio (lipid–MSP1D1)	Incubation temperature
DPPC	90:1	38 °C
DMPC	80:1	25 °C
POPC	65:1	4 °C

4. Add CYP2J2 to yield the desired amount of protein-containing nanodiscs. This amount will need to be amplified, along with the amounts of MSP and lipids, to account for loss in the various filtration and purification steps (*see* **Notes 22** and **23**).
5. Initiate nanodisc self-assembly by adding Bio-Beads SM-2 absorbent or Amberlite XAD-12. Alternatively, detergent may be removed by dialysis against 3 changes of buffer over 24 h. Buffer volume should be 1000-fold greater than the nanodisc mixture volume.
   (a) If Bio-Beads or Amberlite are being used:
      - Clean beads with methanol, followed by thoroughly rinsing with water. Store in H_2O at approximately 1 g/mL.
      - Using a spatula, scoop out a small portion of beads onto a Kimwipes and pat-dry.
      - Add approximately half of the sample volume in beads to the nanodisc mixture (*see* **Note 24**).
      - Seal tube and incubate with rocking for 6–8 h.
      - Filter off beads using a spin column and let the sample incubate overnight with rocking.
6. Concentrate down nanodisc sample to <1 mL and add 10/62 of the sample volume in 72% glycerol in 1 M KPi. Flash-freeze sample and store at −80 °C for at least 4 h (*see* **Note 25**).
7. Fractionate disc samples by injecting on a size-exclusion column in standard 0.1 M KPi buffer (*see* **Note 26**).
8. Collect nanodisc containing fractions and concentrate using an Amicon centrifugal filter.
9. Measure the concentration of the nanodiscs using a UV-Vis spectrophotometer.

3.1.9 Expression and Purification of Rat Cytochrome P450 Reductase (CPR)

Cytochrome P450 reductase (also known as POR, CPR, CYPOR) is a membrane-bound enzyme required for electron transfer from NADPH to cytochrome P450 in the endoplasmic reticulum of the eukaryotic cell. Herein we present the protocol to express *Rattus norvegicus* CPR [7].

1. Transform DH5α cells with the plasmid containing the CPR gene and streak out transformants on an LB plate with ampicillin (20 μg/mL). The streak should aim to produce individual colonies, thus, streak the plate out using a sterile inoculating loop and moving the cells back and forth whilst rotating the plate.
2. Pick a single colony and inoculate a 25 mL starter culture, supplemented with 25 μL 1000× stock ampicillin. Grow overnight at 37 °C, 220 rpm (*see* **Note 27**).
3. Supplement 1 L of autoclaved LB media with 1 mL stock ampicillin and 1 mL of 1 mg/mL freshly made riboflavin [2]. Inoculate with 10 mL of starter culture, and grow at 37 °C and 220 rpm for approximately 4 h (*see* **Note 4**).
4. When the OD_{600} reaches 0.8–1, induce the expression of CPR with the addition of 1 mL 1 mM IPTG. Incubate overnight (18–20 h) at 33 °C and 220 rpm.
5. Harvest cells by centrifugation for 15 min at 2750 × *g*, 4 °C.
   (a) At this point, the cell pellet may be stored at −80 °C, or may be purified directly.
6. Resuspend the cell pellet in cold lysozyme buffer and stir at 4 °C for 30 min (*see* **Note 28**).

Recipe of Lysozyme Buffer (1.2 L buffer per 6 L of culture, adjust if other volume used).

90 mL 1 M Tris–HCl, pH 8 (final 75 mM Tris–HCl).

102.69 g Sucrose (final 250 mM sucrose).

600 μL 0.5 M EDTA (final 0.25 mM EDTA).

24 mg Lysozyme—added fresh (final 0.02 mg/mL lysozyme).

7. Pellet the spheroplasts by centrifugation at 2750 × *g* for 30 min at 4 °C.
   (a) The pellet should have a slimy, mucosal texture.
8. Resuspend pellet in lysis buffer at 4 °C for 30 min. Once homogenized, supplement with fresh DNase and PMSF.

Recipe of Lysis Buffer (120 mL buffer per 6 L of culture, adjust if other volume used).

6 mL 1 M Tris–HCl, pH 8 (final 50 mM Tris–HCl).

120 μL 0.5 M EDTA (final 0.5 mM EDTA).

1.2 mL 100 mM PMSF—made and added fresh (final 1 mM PMSF).

pinch DNase.

9. Sonicate resuspended spheroplasts on ice for 5 cycles, 30 s on/30 s off (*see* **Note 10**).
10. Pellet membrane portion by centrifuging at 104305 × *g* and 4 °C for 45 min. This pellet may be stored at −80 °C.
11. Resuspend pellet in 20 mL column buffer per liter culture at 4 °C, supplemented with 0.5 M NaCl and fresh DTT to 0.1 mM. Once the pellet is resuspended, add Triton to 0.1% v/v.

Recipe of Column Buffer (1 L stock volume, can be adjusted if more is needed).

50 mL 1 M Tris–HCl, pH 8 (final 50 mM Tris–HCl).

250 mL 80% glycerol (final 20% glycerol).

200 μL 0.5 M EDTA (final 0.1 mM EDTA).

M DTT–added fresh (final 0.1 mM).

12. Pellet cellular debris by centrifuging at 104305 × *g* and 4 °C for 1 h.
13. Equilibrate the ADP-agarose column with column buffer supplemented with 0.5 M NaCl, 0.1% Triton, and fresh DTT (0.1 mM).
14. Load supernatant onto column. The supernatant will be viscous and load slowly (*see* **Note 29**).
15. Once loaded, wash column with 50 mL column buffer supplemented with 2.5 mM adenosine, 0.1% Triton, and 0.1 mM fresh DTT.
16. Elute CPR from the column with 100 mL column buffer supplemented with 2 mM $NADP^+$, 0.1% Triton, and fresh DTT.
    (a) Small fractions should be collected until color appears, at which point larger fractions may be collected until color disappears (*see* **Note 30**).
17. Wash the column with column buffer to maintain pH.
18. Concentrate the eluted protein using Amicon centrifugal filters and aliquot. Protein concentration can be measured on a UV-Vis spectrophotometer at 454 nm (*see* **Note 31**).
    (a) Fully oxidized CPR should have a flat baseline from 600 to 700 nm, and no peak at 350 nm. If the protein is not fully oxidized, 10 mM $K_3[Fe(CN)_6]$ may be titrated in 1–10 μL

aliquots until A_{454} reaches a maximum and the dual peak between 350–400 nm collapses to one peak.

3.1.10 Characterization of the Activity of CPR

1. Combine 50 mM cytochrome c and 50 mM NADPH in 0.3 M KPi buffer (pH 7.4) and equilibrate to 30 °C.
2. Add 2 mM CPR to a final volume of 1 mL.
3. Measure cytochrome c reduction at 550 nm for 10 min ($\varepsilon = 2.1 \times 10^4\ M^{-1}\ cm^{-1}$).
   (a) Activity should be ~50 μM cyt c/min/mg CPR.

3.2 Materials and Methods for CYP725A4

3.2.1 Introduction

Taxadiene-5α-Hydroxylase (CYP725A4) is a membrane-bound plant CYP that catalyzes the oxidation of taxadiene to taxadiene-5α-ol, which is the first step of taxol synthesis. Herein, we provide the protocols for the bacterial expression and purification of the CYP725A4 chimera with taxus CPR. The redox partner of CYP725A4 is *Taxus cuspidata* CPR. Furthermore, we study the binding of taxadiene to the CYP725A4 and incorporate into nanoscale lipid bilayers (nanodiscs) and measure the taxadiene conversion to both taxadiene-5α-ol and OCT [6].

3.2.2 Expression of CYP725A4-Chimera with Taxus cuspidata CPR

1. Streak out a plate of CYP-CPR strain on an LB and ampicillin agar plate, incubate at 37 °C overnight. Streak out the plate in order to create individual colonies. Streak the plate by using an inoculation loop, and streaking the bacteria out whilst moving the plate around.
2. Culture a single colony with the CYP725A4 gene in 25 mL of LB broth. Pick the culture with the disposable tip of a micropipette under a flame, and drop the pipette tip into the flask. Add 60 μg/mL of Spectinomycin or 0.1 mg/mL of ampicillin under a flame. Grow overnight at 30 °C at 250 rpm (*see* **Note 2**).
   (a) Ampicillin is reserved for unlinked constructs and spectinomycin is for linked constructs.
3. Prepare 3 L of TB as follows:
   (a) 36 g tryptone.
   (b) 72 g yeast extract.
   (c) 6 g peptone.
   (d) 300 mL of 1 M KP_i at pH 7.4.
   (e) 12 mL of 80% glycerol.
4. Pour 500 mL of TB into six separate 2.8 L Fernbach flasks, cover with aluminum foil and autoclave.
5. After the flasks cool, add the following to each flask under a flame, and flaming the lip of the flasks:
   (a) 500 μL of spectinomycin.
   (b) 0.125 mL of trace minerals.

- Trace mineral composition is: 166 mM $FeCl_3$, 15 mM $ZnCl_2$, 8.3 mM $NaMoO_4$, 8.4 mM $CoCl_2$, 9 mM $CaCl_2$, 7.3 mM $CuCl_2$, and 8.1 mM H_3BO_3 in 1.6 M HCl.

(c) 500 μL of riboflavin (stock: 1 mg/1 mL H_2O.

(d) 6.725 μL of 1 M thiamine (make sure the thiamine is filter-sterilized).

6. Inoculate each flask under a flame with 5 mL of the starter culture (*see* **Note 4**).
7. Grow the flasks at 250 rpm at 30 °C for 2.5 h in a shaking incubator.
8. After the 2.5 h add 0.25 mL of δ-ALA (stock: 0.838 g/10 mL H_2O) to each flask. Continue to grow for about 2 more hours at 250 rpm.
9. After the cells reach an OD_{600} of 0.9, induce the flasks by inoculating 0.5 mL of 1 mM of IPTG stock and 2 g of arabinose into each flask.
10. Grow for an additional 44–48 h at 22 °C and 250 rpm after induction.
11. After the cells have finished growing, centrifuge the 3 L of media at 11000 × *g* for 30 min at 4 °C. Discard the supernatant and scrape out the pellet. Place the pellet into a freezer bag and store in −80 °C.

3.2.3 Purification of CYP725A4-Chimera with Taxus cuspidata *CPR*

1. Resuspend the pellet in 500 mL of a lysozyme buffer at 4 °C for 30 min. The buffer is prepared as follows:

(a) 75 mM Tris–HCl pH 8.

(b) 0.25 M sucrose.

(c) 0.25 M EDTA.

(d) 0.02 mg/mL lysozyme.

2. Pellet the solution at 11000 × *g* for 30 min and resuspend in the lysis buffer. The buffer is prepared as follows:

(a) 0.1 M pH KP_i Buffer at pH 7.4.

(b) 20% glycerol.

(c) 6 mM $MgCl_2$.

(d) 0.1 mM DTT.

(e) 0.2 mM PMSF.

(f) 5 mg of DNase.

3. Sonicate the cells in the lysis buffer on ice, in intervals of 40 s on, 40 s rest 6 times each for every 50 mL (*see* **Note 10**).
4. Centrifuge the lysate at 4 °C for 45 min at 142,000 × *g*.
5. Suspend the membrane fraction in the extraction buffer for 5–6 h at 4 °C. The extraction buffer is prepared as follows:
   (a) 0.1 M pH 7.4 KP_i.
   (b) 20% glycerol.
   (c) 200 mM NaCl.
   (d) 0.1 mM DTT.
   (e) 1% cholate.
6. After resuspension, centrifuge the mixture at 140,000 × *g* for 45 min at 4 °C.
7. Load the supernatant on a Ni-NTA column that has been preincubated with a column buffer. The column buffer was prepared as follows:
   (a) 0.1 M KP_i pH 7.4.
   (b) 20% glycerol.
   (c) 0.1% cholate.
   (d) 1 mM EDTA.
   (e) 1 mM DTT.
8. Wash the loaded column with 50 mL of column buffer.
9. Wash the loaded column with 50 mL of column buffer +20 mM imidazole.
10. Elute the protein with the 100 mL of column buffer +200 mM imidazole and 200 mM NaCl.
11. Concentrate CYP725A4 down by centrifugation with an Amicon centrifugal filter.
12. Buffer exchange the protein by concentrating the protein through centrifugation in a concentrator and gradual addition of the exchange buffer. The exchange buffer is created as follows:
    (a) 0.1 M pH 7.4 KP_i.
    (b) 20% glycerol.
    (c) 0.1% cholate.
    (d) 1 mM DTT.
    (e) 200 mM NaCl.

13. Measure concentration by UV-Vis spectroscopy using a 1 cm quartz cuvette at 417 nm. Use an extinction coefficient of 116,000 M^{-1} cm^{-1}.

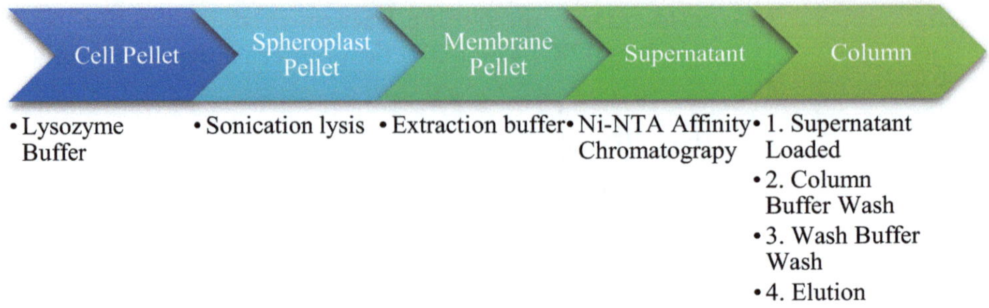

Schematic depicts the CYP725A4-CPR chimera extraction process.

3.2.4 Expression and Purification of Stevia CPR

1. Inoculate 2–25 mL LB media aliquots with 1 colony of truncated Stevia CPR (SrCPR) each and 25 μL 1000× ampicillin. Grow overnight at 37 °C and 220 rpm.
2. Prepare six flasks with 500 mL Terrific Broth (TB) and supplement with 1 mg/L riboflavin, mineral spike, and 0.5 mL 1000× ampicillin after autoclaving. Before beginning culture, warm flasks to 37 °C.
3. Inoculate each flask of TB with 5 mL of the overnight starter culture (*see* **Note 4**) and grow at 32 °C and 180 rpm until the OD_{600} reaches 0.6.
4. Induce with 0.5 mL 1 mM IPTG and grow overnight at 32 °C and 180 rpm.
5. Harvest cells by centrifugation at 4 °C at 24515 × *g* for 15 min.
6. Resuspend cell pellet for 30 min in 100 mL of lysozyme buffer (*see* previous protocol for *R. norvegicus* CPR) with 1 mg lysozyme per liter of culture (*see* **Note 28**).
   (a) All buffers for Stevia CPR purification should be assumed to be the same as for RnCPR purification unless otherwise specified.
7. Pellet the spheroplasts by centrifugation at 4 °C at 2750 × *g* for 30 min.
8. Resuspend the pellet at 4 °C in 10 mL lysis buffer per liter of culture.
9. Sonicate resuspended spheroplasts on ice for 5 cycles, 30 s on/30 s off.
10. Pellet membrane portion by centrifuging at 104305 × *g* and 4 °C for 45 min. This pellet may be stored at −80 °C.

11. Resuspend the membrane fraction for 1 h in 10 mL of column buffer per liter of culture supplemented with 0.5 M NaCl and 0.1 mM fresh DTT.
12. After 1 h, add 1% (v/v) cholate and 0.5% (v/v) tergitol, and incubate another hour.
13. Centrifuge at 4 °C and 104305 × *g* for 1 h to pellet the cellular debris.
14. Equilibrate the ADP-Agarose column with column buffer supplemented with 300 mM NaCl, 0.1% tergitol, and 0.5% cholate.
15. Load the supernatant onto the equilibrated column and wash with column buffer supplemented with 1.5 μM riboflavin-5′-phosphate, and 2.5 mM adenosine.
16. Elute protein from the column using column buffer supplemented with 500 mM NaCl, 0.2% tergitol, 0.2% cholate, 2 mM NADPH, and 0.1 mM fresh DTT.
17. Concentration can be determined on a UV-Vis spectrophotometer at 456 nm.

3.2.5 Incorporation of CYP725A4 into Nanodiscs and Activity Assay

Materials are same as Subheading 3.1.8.

1. Assembly of CYP725A4 Nanodiscs: Solubilize POPC in cholate as detailed previously.
2. Add MSP1E3D1 and incubate at 4 °C for 1 h.
3. Add CYP725A4 or linked CYP725A4-CPR to the P450–MSP mixture in a 1:15 ratio and incubate at 4 °C for 1 h.
4. Add Amberlite XAD-12 beads to remove detergent as detailed previously.
5. Purify nanodiscs on a Ni-NTA column, eluting with 0.1 M KPi buffer supplemented with 0.2 M imidazole.
6. Concentrate nanodiscs and determine the concentration using a UV-Vis spectrophotometer.

3.2.6 Activity Assay of CYP725A4 in Nanodiscs

1. Turn on a UV-Vis spectrophotometer at least 15 min before use. Set the spectrophotometer to maintain a temperature of 32 °C.
2. Blank the spectrophotometer using 0.25 M Tris–HCl buffer.
3. In a clean quartz cuvette, combine 50 pmol CYP2725A4, 150 pmol SnCPR or RnCPR, Tris–HCl buffer (final 0.25 M Tris–HCl), 200 μM NADPH, 2.5 μM FMN, 2.5 μM FAD, 50 μM taxadiene, 2 mM glucose-6-phosphate, 2 units glucose-6-phosphate dehydrogenase, and MilliQ H_2O.
   (a) Controls and SnCPR reactions were performed in 2 mL, RnCPR reactions were performed in 1 mL.

4. Allow reaction to incubate for 1 h at 32 °C.
5. At the 1-h mark, quench reactions with an equal volume of 4 M NaCl.
6. Extract the organic layer quenched reaction twice with 2× the reaction volume of 4:1 hexane–ether.
7. Dry the organic layer under a stream of ultrapure nitrogen, then resuspend in 100 μL hexanes for further analysis by GC-MS.

4 Notes

1. For best results, streak plate in four sections with different dilutions, sterilizing the loop in between each, and running the loop through a small portion of the previous section to grab cells.
2. As you will need 30 mL to seed a full 3 L growth, it is best to make three overnight cultures.
3. For best results, flame both the inside of the foil cap and the entire circumference of the flask lip immediately after opening and before closing. The flask lip should also be flamed in between additions of different ingredients.
4. If the volume of one overnight culture is insufficient to seed all flasks, combine with a second overnight culture before seeding any flasks. The cultures should all be seeded with the same mix of starter culture.
5. The OD_{600} of the flasks at this point should be ~0.8–1.2.
6. As with any midculture addition step, flasks should all be removed from the incubator and replaced at the same time to ensure homogeneity of the culture. Flasks may be left at room temperature until all flasks in the set are prepared.
7. The OD_{600} at this point may be anywhere between 1.0 and 1.2. In general inducing at lower OD will reduce overall yield.
8. The pellet should have a red tint to indicate that the cytochrome P450 has been expressed.
9. The pellet may be homogenized to speed up suspension. If taken from the freezer, the pellet should be allowed to thaw while stirring in the resuspension buffer before homogenization.
10. Swirl the solution in the beaker while homogenizing for efficient lysis.
11. The supernatant should have a red color.
12. Once supernatant has loaded, the column should appear brown/red.

13. Collect the flow-through from loading the CYP2J2 supernatant. If the column volume is small, or expression was very robust, there may be excess CYP2J2 that does not bind the column during the initial loading. After elution of the first round of protein, the column may be reequilibrated and this flow-through loaded onto the column for a second purification.
14. Concentration and buffer exchange should take place at 4 °C, at speeds no higher than 2050 × *g* and depends on the concentrators used. CYP2J2 tends to settle down during centrifugation and there is concentration gradient. Therefore the Amicon tubes should be inverted 3–5 times after each spin to prevent aggregate formation.
15. A glass stir bar must be used as Teflon stir bars will occlude O_2.
16. When purging the protein, the stream of inert gas must be as gentle as possible to prevent foaming of the protein. Immediately after purging, needle holes should be sealed with nail polish to prevent oxygen from reentering the system.
17. Approximately one spatula-tip-full of dithionite should be added to the 3 mL of anaerobic KPi. This does not need to be precise.
18. Mixing should take place as quickly as possible (in a few seconds), without forming bubbles.
19. Avoid coming into contact with the interstice between the aqueous and organic layers as this could introduce column-clogging contamination.
20. When drying on the rotovap, different lipids will produce different films on the tube wall. For example, POPC leaves a clear film, but the introduction of any percentage POPS leaves an opaque white film. Also, insufficient drying on the rotovap before vacuum desiccator storage will cause the lipids to form an insoluble mass that may not be soluble later even with sonication.
21. If the discs are to be made with the enzyme then the MSP incubation step should be increased to 2 h.
22. You can use either buffer exchanged CYP2J2 or imidazole bound CYPJ2 to prepare nanodiscs.
23. Common yield for this nanodisc procedure with the target membrane protein is between 40 and 60%. The yield varies depending on the type of membrane protein that is incorporated.
24. Once Bio-Beads or Amberlite has been added, the solution should still be reasonably fluid so that it can be stirred or rocked to ensure efficient removal of detergent.

25. Nanodisc may be frozen for optimal efficiency.
26. The HPLC chromatogram typically shows three peaks—a small peak of aggregates, a single large peak that corresponds to the well-formed nanodiscs, and another peak of protein aggregates.
27. If a full 6 L of culture is being grown, start at least three overnight cultures of CPR.
28. Cell pellet must be stirred 30 min AFTER full resuspension is achieved. A homogenizer may be used, as this pellet does not resuspend quickly.
29. To speed up column loading, you can use gravity and suspend the loading solution multiple feet higher than the resin column.
30. The color of the eluted protein can range from blue-green to banana yellow depending on oxidation state of the CPR. This may not be as obvious until after concentration.
31. If the CPR is not needed for some time, it can be frozen and will oxidize naturally. Fully oxidized frozen CPR should be bright yellow.

References

1. Zelasko S, Palaria A, Das A (2013) Optimizations to achieve high-level expression of cytochrome P450 proteins using Escherichia coli expression systems. Protein Expr Purif 92 (1):77–87. https://doi.org/10.1016/j.pep.2013.07.017 Epub 2013/08/27. [pii]. S1046-5928(13)00143-5
2. McDougle DR, Palaria A, Magnetta E, Meling DD, Das A (2013) Functional studies of N-terminally modified CYP2J2 epoxygenase in model lipid bilayers. Protein Sci 22(7):964–979. Epub 2013/05/11. PubMed PMID: 23661295; PMCID: 3719090. https://doi.org/10.1002/pro.2280
3. McDougle DR, Baylon JL, Meling DD, Kambalyal A, Grinkova YV, Hammernik J, Tajkhorshid E, Das A (2015) Incorporation of charged residues in the CYP2J2 F-G loop disrupts CYP2J2-lipid bilayer interactions. Biochim Biophys Acta 1848(10 Pt A):2460–2470. https://doi.org/10.1016/j.bbamem.2015.07.015 Epub 2015/08/02. [pii]. S0005-2736(15)00237-0
4. von Wachenfeldt C, Richardson TH, Cosme J, Johnson EF (1997) Microsomal P450 2C3 is expressed as a soluble dimer in Escherichia coli following modifications of its N-terminus. Arch Biochem Biophys 339(1):107–114. PubMed PMID: ISI:A1997WL10900015. https://doi.org/10.1006/abbi.1996.9859
5. Davydov DR, Davydova NY, Sineva EV, Kufareva I, Halpert JR (2013) Pivotal role of P450-P450 interactions in CYP3A4 allostery: the case of alpha-naphthoflavone. Biochem J 453:219–230 PubMed PMID: ISI:000321808800007
6. Biggs BW, Rouck JE, Kambalyal A, Arnold W, Lim CG, De Mey M, O'Neil-Johnson M, Starks CM, Das A, Ajikumar PK (2016) Orthognal assays clarify the oxidative biochemistry of taxol P450 CYP725A4. ACS Chem Biol 11 (5):1445–1451 PubMed PMID: ISI:000376473600034
7. Wang M, Roberts DL, Paschke R, Shea TM, Masters BSS, Kim JJP (1997) Three-dimensional structure of NADPH-cytochrome P450 reductase: prototype for FMN- and FAD-containing enzymes. Proc Natl Acad Sci U S A 94(16):8411–8416. PubMed PMID: ISI:A1997XQ12400023. https://doi.org/10.1073/pnas.94.16.8411

Chapter 6

Delta Integration CRISPR-Cas (Di-CRISPR) in *Saccharomyces cerevisiae*

Shuobo Shi, Youyun Liang, Ee Lui Ang, and Huimin Zhao

Abstract

Despite the advances made in genetic engineering of *Saccharomyces cerevisiae*, the multicopy genomic integration of large biochemical pathways remains a challenge. Here, we developed a Di-CRISPR (delta integration CRISPR-Cas) platform based on cleavage of the delta sites by Clustered Regularly Interspaced Short Palindromic Repeats (CRISPR) and CRISPR-associated systems (Cas) to enable unprecedented high-efficiency, multicopy, markerless integrations of large biochemical pathways into the *S. cerevisiae* genome. Detailed protocols are provided on the entire workflow which includes pDi-CRISPR plasmid and donor DNA construction, Di-CRISPR-mediated integration and analysis of integration efficiencies and copy numbers through flow cytometry and quantitative polymerase chain reaction (qPCR).

Key words Delta integration, CRISPR-Cas, *Saccharomyces cerevisiae*, Genome integration, Genome engineering, Synthetic biology

1 Introduction

Owing to its many advantages such as genetic tractability, tolerance toward harsh conditions and inhibitors, and sound history of industrial use, *S. cerevisiae* has been widely adopted as a production host for fuels, pharmaceuticals, and other value-added chemicals [1, 2]. For these industrial applications, the genomic integration of heterologous genes and biochemical pathways through homologous recombination (HR) confers greater stability and is generally preferred over the use of episomal plasmids [3, 4]. However, the multicopy integration of large multigene pathways is nontrivial and has been confounded by low efficiencies and tedious multistep workflows [5, 6], with integration efficiencies and copy numbers declining rapidly with larger donor DNA [7, 8]. To increase HR efficiencies, different nucleases such as homing endonucleases,

Shuobo Shi and Youyun Liang contributed equally to this work.

Christine Nicole S. Santos and Parayil Kumaran Ajikumar (eds.), *Microbial Metabolic Engineering: Methods and Protocols*, Methods in Molecular Biology, vol. 1927, https://doi.org/10.1007/978-1-4939-9142-6_6,

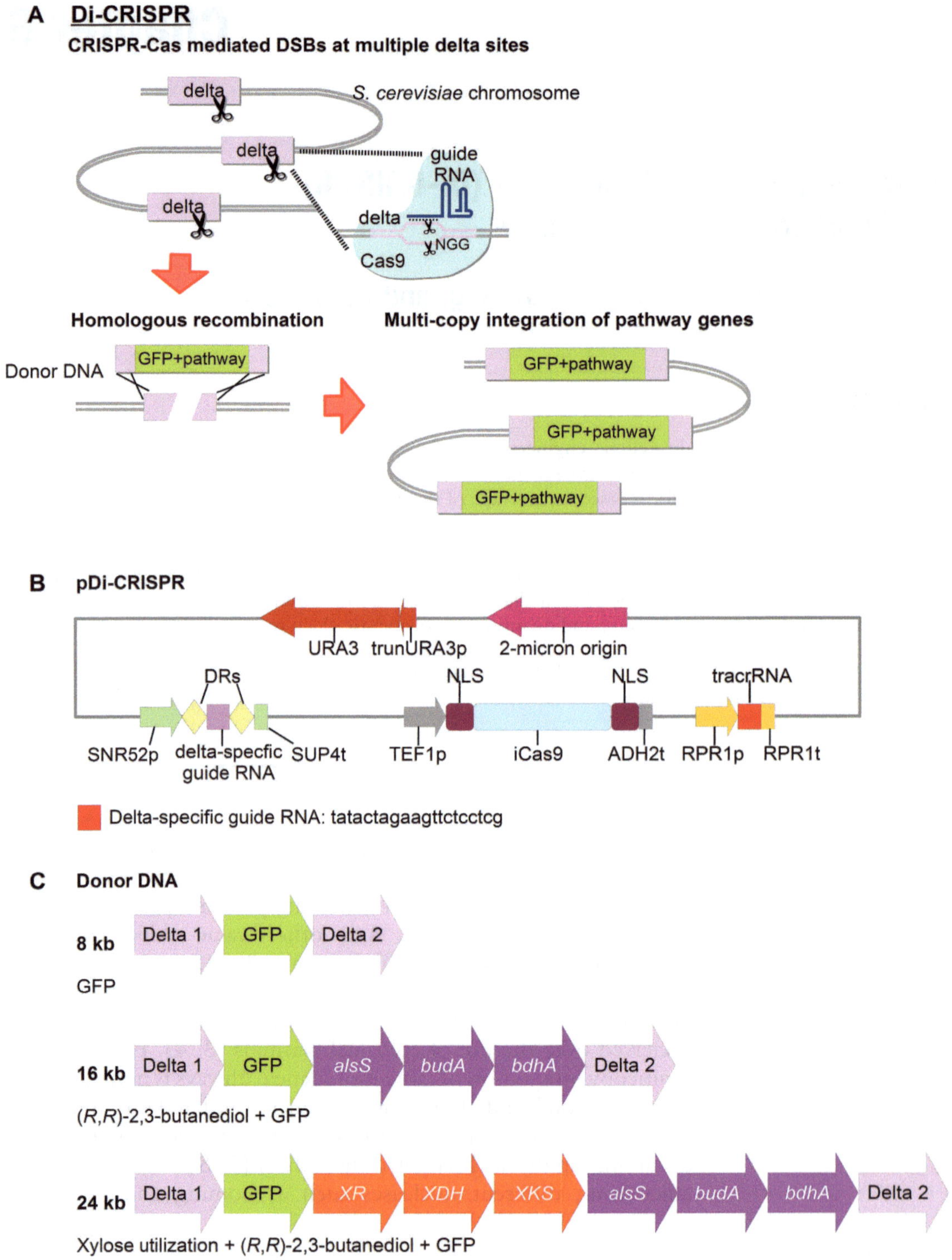

Fig. 1 Schematic representation of Di-CRISPR. (**a**) CRISPR-Cas was used to induce multiple DSBs at the delta sites. The DSBs helped to enhance homologous recombination of biochemical pathways at the cut sites. (**b**) The pDi-CRISPR plasmid was modified from pCRCT [13] described by Bao et al. by replacing the original guide

Transcription Activator-Like Effector Nucleases (TALEN) [9, 10] and CRISPR-Cas have been used to generate double stranded breaks (DSBs) at the target integration sites [9–13]. While these approaches effectively boosted integration efficiencies, the simultaneous integration of multiple large biochemical pathways, which is crucial for rapid strain construction and engineering, had not been attempted.

To address this void, we developed Di-CRISPR (delta integration CRISPR-Cas) to facilitate high-efficiency, multicopy, markerless integrations of large biochemical pathways into the *S. cerevisiae* genome (Fig. 1) [14]. In this platform, the highly homologous, noncoding Ty retrotransposon delta sites distributed throughout the yeast genome are used as landing pads for integration of large biochemical pathways. To mediate high-efficiency integration, multiple DSBs are simultaneously generated at selected delta sites in the presence of compatible donor DNA (Fig. 1a). The DSBs are generated with the help of CRISPR-Cas through a specially designed delta-specific guide RNA. This guide RNA and the Cas9 protein are housed on a single ultrahigh copy pDi-CRISPR plasmid (Fig. 1b) [14], designed in accordance to the previously reported HI-CRISPR system [13]. The linear donor DNA, harboring the pathway of interest, contains the two halves of the delta sequence at each end to facilitate accurate and efficient HR at the generated DSBs (Fig. 1b). In this work, 8 kb, 16 kb, and 24 kb donor DNA containing GFP (Fig. 1c) were constructed to provide an easy readout of the relative integration efficiencies and copy numbers. Following DNA construction, the Di-CRISPR plasmid and donor DNA were cotransformed in *S. cerevisiae* to induce donor DNA integration at the delta sites. After which, the engineered strains generated with our Di-CRISPR platform were then quickly analyzed through flow cytometry and quantitative polymerase chain reaction (qPCR) to yield efficiencies and copy numbers. Using our platform, we were able to achieve the unprecedented 18-copy integration of a 24 kb model pathway for xylose utilization and (R,R)-2,3-butanediol (BDO) production in a single step. With the high efficiency and simplicity, we envision that Di-CRISPR will be extremely useful for genomic integrations in *S. cerevisiae*.

Fig. 1 (continued) sequence array with a delta-specific guide RNA. (**c**) Linear donor DNA of 8 kb, 16 kb, and 24 kb were constructed by linearizing plasmid DNA, pSS99, pSS100, and pSS101, and integrated using the Di-CRISPR system. GFP was included in the donor DNA to facilitate analysis with flow cytometry. A functional butanediol (BDO) production pathway (*alsS, budA, bdhA*) was included in the 16-kb construct, while the 24-kb construct contained xylose utilization (*XR, XDH, XKS*) and BDO production pathways

2 Materials

2.1 Strains

1. DH5α competent *Escherichia coli*.
2. Laboratory strain *S. cerevisiae* HZ848 (*MATα*, *ade2-1*, *ura3Δ*, *his3-11*, *15*, *trp1-1*, *leu2-3*, *112*, and *can1-100*) [15].

2.2 Plasmids and DNA

1. pCRCT plasmid (available from Addgene) [13].
2. pDi-CRISPR plasmid (pDi-CRISPR with guide RNA TATACTAGAAGTTCTCCTCG is available from Addgene. Alternatively, a new pDi-CRISPR with a distinct guide RNA sequence can be constructed) [14].
3. Plasmid pRS426-TEF1p-GFP-TEF1t [16], plasmid pRS423 (NEB), pRS416 (NEB), plasmid pRS414-ADH1p-ADH1t [17], plasmid pRS414-PGK1p-CYC1t [17], plasmid pRS414-PYK1p-ADH2t [17].
4. *Bacillus subtilis* genomic DNA (ATCC, Manassas, VA, USA), *Enterobacter aerogenes* genomic DNA (ATCC, Manassas, VA, USA).

2.3 Cell Culture Media and Substrates

1. S.O.C. medium (Invitrogen, Carlsbad, CA, USA).
2. Luria broth (LB) medium: dissolve 12 g of LB Broth into 600 mL Milli-Q water. Autoclave at 121 °C for 20 min.
3. 1000× ampicillin (Amp) stock: dissolve 1 g of ampicillin sodium salt into 10 mL Milli-Q water and filter through 0.22 μm filter. Aliquot is stored at −20 °C. For LB + Amp medium, dilute ampicillin stock 1000-fold to a working concentration of 100 μg/mL.
4. Bacto with Amp (LB + Amp) Agar: Supplement 20 g/L agar into proper liquid medium before autoclave. Cool the autoclaved mixture to 50–60 °C, add 600 μL of above Amp stock to a working concentration of 100 μg/mL. Pour 20 mL of the molten agar into each 100 × 10 mm petri dish and allow to solidify.
5. YPD medium: dissolve 6 g of yeast extract, 12 g of peptone, 12 g of dextrose, and 60 mg of adenine hemisulfate in 600 mL of Milli-Q water. Autoclave at 121 °C for 20 min.
6. SC medium (Formedium, Norfolk, UK).

2.4 Molecular Biology Enzymes and Reagents

1. *Avr*II, *BamH*I, *Bsa*I, *Bsp120*I, *Cla*I, *EcoR*I, *Sac*II, and *Sal*I restriction enzymes.
2. TAE electrophoresis buffer: Dilute the 50× TAE electrophoresis buffer with Milli-Q water to a 1× working concentration of 40 mM Tris, 20 mM acetic acid, and 1 mM EDTA.

3. 0.7% agarose gel in 1× TAE buffer: Dissolve 0.7 g of agarose into 100 mL of 1× TAE buffer by heating in the microwave. Cool the solution to approximately 70–80 °C. Add 10 μL of SybrSafe (Invitrogen) into the solution and mix well. Pour 50 mL of solution onto an agarose gel rack with appropriate 2- or 8-well combs. For 0.7% and 1.0% agarose gel, add 0.7 g and 1.0 g of agarose into 100 mL of 1× TAE buffer.
4. Alkaline phosphatase, calf intestinal (CIP).
5. 6× Loading dye.
6. T4 polynucleotide kinase (T4 PNK).
7. T4 DNA Ligase (400,000 units/mL).
8. 1.0 M NaCl.
9. Thialysine (S-2-aminoethyl-L-cysteine).
10. QIAprep Miniprep Kit.
11. QIAquick PCR Purification and Gel Extraction Kit.
12. Phosphate buffered saline (PBS).
13. Wizard Genomic DNA Purification Kit.
14. KOD DNA polymerase (Merck, Darmstadt, Germany).
15. 100 mM dATP, dGTP, dCTP, and dTTP.
16. 70% (v/v) ethanol.
17. 30% (v/v) glycerol solution: mix 30 mL glycerol with 60 mL ddH_2O, and bring the total volume to 100 mL using ddH_2O.- Filter through 0.22 μm filter.
18. Gibson Assembly Cloning Kit (NEB).
19. UltraPure Salmon Sperm DNA Solution (ssDNA) (Stratagene, San Diego, CA, USA).
20. 1.0 M Lithium acetate (1.0 M LiAc): dissolve 10.2 g of lithium acetate dihydrate in 100 mL of ddH_2O. Autoclave at 121 °C for 20 min.
21. LiAc/TE: Mix 5 mL 1.0 M sterile LiAc with 0.5 mL sterile 1 M Tris–HCl, pH 7.5 and 0.1 mL sterile 500 mM EDTA. Top up to 50 mL with sterile ddH_2O.
22. 50% (w/v) PEG (Mw 3350): Dissolve 25 g PEG in 40 mL water by incubating in a 50 °C water bath. Top up the solution to 50 mL. Autoclave at 121 °C for 20 min.
23. FastStart Essential DNA Green Master (Roche, Basel, Switzerland).

2.5 Equipment and Software

1. NanoDrop2000c: measure DNA concentrations.
2. Horizontal electrophoresis system: Mini-Sub Cell GT cells and PowerPac Basic 300 V Power Supply.

3. PCR thermal cycler.
4. Molecular Imager Gel Doc system.
5. Water bath.
6. Microcentrifuge.
7. Shaking incubator.
8. Incubator.
9. Benchtop centrifuge.
10. Spectrophotometer.
11. Heating block.
12. Vortex mixer.
13. Becton Dickson LSR II flow cytometer.
14. SZX16 fluorescent stereo microscope (Olympus, Tokyo, Japan).
15. Roche Light Cycler® 96 System.
16. LightCycler® 480 Multiwell Plate 96.

2.6 Software

1. GGGenome search engine (https://gggenome.dbcls.jp/sacCer3/10/).
2. BD FACSDiva Software (BD Biosciences, San Jose, CA, USA).
3. Primer 3 Version 0.4.0 (http://bioinfo.ut.ee/primer3-0.4.0/primer3/).

3 Methods

3.1 Di-CRISPR Plasmid Design and Construction

1. Design delta-specific guide RNA. Select a unique 20 bp region flanked by a protospacer adjacent motif (PAM) with the sequence NGG in the delta sequence (*see* **Note 1**). Run the 23 nucleotides (unique 20 bp + NGG) through the GGGenome search engine using *S. cerevisiae* reference S288C genome to find out how many DSBs will be generated. Ensure that DSBs are only generated within the delta sites. In our work the guide RNA TATACTAGAAGTTCTCCTCG (*see* **Note 2**) has been selected.
2. Digest pCRCT plasmid [13] (Fig. 2a) at 37 °C for 3 h in a thermocycler. Reaction condition: 2 μL of *Bsa*I, 5 μL of 10× CutSmart buffer, and 1 μg of pCRCT. Add ddH_2O to a final volume of 50 μL. After the reaction, add 1 μL of CIP and incubate at 37 °C for 1 h. Add 10 μL of 6× loading dye to the digestion product and load the sample into 0.7% agarose gels made with 2-well combs and perform gel electrophoresis at 120 V for 40 min. Image the gel with Molecular Imager Gel Doc system and cut the gel slice containing linearized pCRCT

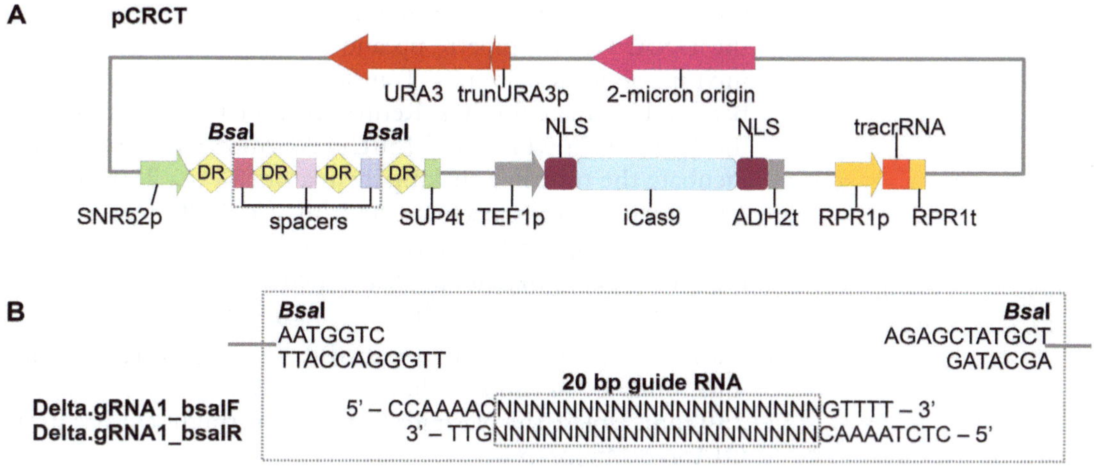

Fig. 2 Plasmid construction for Di-CRISPR. (**a**) Guide array of pCRCT plasmid [13] can be excised by *Bsa*I. (**b**) Delta-specific guide RNA. Oligonucleotides encode unique delta-specific 20-bp guide RNA and complementary ends to *Bsa*I-digested pCRCT

(10,346 bp). Purify using QIAquick Gel Extraction Kit and measure the concentrations using NanoDrop.

3. Generate single stranded oligonucleotides encoding guide RNA and *Bsa*I sticky ends. Design and order single stranded oligonucleotides delta.gRNA1_bsaIF and delta.gRNA1_bsaIR (Fig. 2b). delta.gRNA1_bsaIF comprises of CCAAAAC (unique 20 bp)GTTTT whereas delta.gRNA1_bsaIR comprises of CTCTAAAAC(reverse complement of unique 20 bp)GTT (*see* **Note 3**).
4. Phosphorylate single stranded oligonucleotides. Phosphorylate delta.gRNA1_bsaIF and delta.gRNA1_bsaIR at 37 °C for 1 h in a 50 μL reaction containing 1 μL delta.gRNA1_bsaIF (100 μM), 1 μL delta.gRNA1_bsaIR (100 μM), 5 μL 10× T4 PNK buffer, and 1 μL T4 PNK.
5. Anneal phosphorylated oligonucleotides. Add 2.5 μL of 1 M NaCl to the above reaction containing phosphorylated oligonucleotides. Using a thermocycler, incubate for 5 min at 95 °C and slowly cool down to room temperature (1 °C/min). Dilute the annealed double stranded oligonucleotide by 10×.
6. Ligate guide RNA oligonucleotide and *Bsa*I-digested pCRCT at 25 °C for 1 h to generate pDi-CRISPR. The ligation reaction contains 30 ng of *Bsa*I-digested pCRCT, 0.5 μL of annealed and diluted guide RNA oligonucleotide, 1 μL of 10× T4 Ligase buffer, 0.5 μL of T4 DNA Ligase in a 10 μL reaction.
7. Transform 10 μL of ligation products into 50 μL chemical competent *E. coli* DH5α. For the transformation, thaw frozen DH5α cells in a 14 mL conical tube on ice and add ligation

products slowly, gently swirling the pipette tip while dispensing the DNA. Do not pipette up and down to mix. Incubate the mixture on ice for 30 min. Transfer the tube to a 42 °C water bath and incubate for 30 s. Return the tube to ice and incubate for 2 min. After which, add 1 mL prewarmed S.O.C media and incubate the tube in a shaking incubator (37 °C, 250 rpm) for 1 h. Spin down cells in a microcentrifuge at 3000 × *g* for 5 min, remove the supernatant and resuspend cells in 100 μL S.O.C medium. Spread the 100 μL cell suspension on LB + Amp plate and allow to grow overnight (16–20 h) at 37 °C.

8. Plasmid preparation for pDi-CRISPR. Inoculate three colonies into 5 mL LB + Amp medium and grow overnight at 37 °C, 250 rpm. Isolate the plasmids from the overnight culture using QIAprep Miniprep Kit.

3.2 Donor DNA Construction

Three donor DNA fragments with different sizes were constructed by linearizing plasmid DNA, pSS99, pSS100, and pSS101 (Fig. 1c). Plasmids were constructed by the following procedures:

1. Inoculate a single colony of the HZ848 strain into 3 mL YPD medium and grow overnight at 30 °C and 250 rpm (*see* **Note 4**). Isolate the genomic DNA from 3 mL cell culture using Wizard Genomic DNA Purification Kit in six separate reactions. Measure the concentrations using NanoDrop.
2. Amplify the delta fragments from the isolated genomic DNA by PCR. Two DNA fragments, Delta1 and Delta2, each containing delta-sequence elements were PCR amplified from *S. cerevisiae* HZ848 genomic DNA. Primers for Delta1 fragment (Delta1For1 and Delta1Rev1) and Delta2 fragment (Delta2For1 and Delta2Rev1) can be found in Table 1. The PCR reaction conditions were set according to the manual of KOD DNA Polymerase. Purify the PCR products using QIAquick PCR Purification Kit and measure the concentrations using NanoDrop.
3. Amplify the GFP expression cassette 1, *TEF1p-GFP-TEF1t*, from plasmid pRS426-TEP1p-GFP-TEF1t by PCR with upstream-primer (TEF1pFor1) and downstream-primer (TEF1tRev1). The PCR reaction condition was set according to the manual of KOD DNA Polymerase. Purify the PCR product using QIAquick PCR Purification Kit and measure the concentrations using NanoDrop.
4. Digest plasmid pRS423 at 37 °C for 6 h in a thermocycler. Reaction condition: 1 μL of *Cla*I, 1 μL of *Eco*RI, 5 μL of 10× CutSmart buffer, and 1 μg of pRS423. Add ddH_2O to a final volume of 50 μL. Purify the digestion product using QIAquick

Table 1
The primers used in this work

No	ID	Primer name	Primer sequence (5′ → 3′)
Primers for DNA manipulation			
1	Delta-specific guide RNA	delta. gRNA1_bsaIF	CCAAAACTATACTAGAAGTTCTCCTCGGTTT
		delta. gRNA1_bsaIR	CTCTAAAACCGAGGAGAACTTCTAGTATAGTT
2	Delta1	Delta1For1	CTCGAGGTCGACGGTATCGATACTCGAGGATATAGGAATCCTCAAAAT
		Delta1Rev1	GAGCTCGTTTAAACGGGCCTAGGTGTTGGAATAGAAATCAACTATCATC
3	Delta2	Delta2For1	CCTAGGCCCGTTTAAACGAGCTCTGGAAGCTGAAACGTCTAACG
		Delta2Rev1	AAAAAAGGAGTAGAAACATTTTGAAGCTATGAGAACTTCTAGTATATTCTGTATACCTAATATTATAG
4	GFP expression cassette 1	TEF1pFor1	GGTATACAGAATATACTAGAAGTTCTCATAGCTTCAAAATGTTTCTACTCC
		TEF1tRev1	CTAGTGGATCCCCCGGGCTGCAGGATAGCGCCGATCAAAGTATTTGTTAC
5	GFP expression cassette 2	TEF1pFor2	CAAGACAAAGAAATTGCATTGTCTATAACGATAACAAAAGACATCGATAGCTTCAAAATGTTTCTACTCC
		TEF1tRev2	CTATTTTGAAGGCATGCAAGAGGTTCTGTGACTATGCAACAGCAGATAGCGCCGATCAAAGTATTTG
6	BDO expression cassette	BDO-For1	GAGCGCGCGTAATACGACTCACTATAGGGCGAATTGGGTAC AGCTACTAATAGGATAAATTATAGG
		BDO-Rev1	GATTCCATTTTGAGGATTCCTATATCCTCGAGTATCGATACCGTGCCTGCAGGTCGAGATC
7	Xylose utilization cassette	Xylose-For1	GTAACAAATACTTTGATCGGCGCTATCCTGCAGCCCGGGGTGCCTGCAGGTCGAGATC
		Xylose-Rev1	CTCACTAAAGGGAACAAAAGCTGGAGCTCCACCGCGGTGGCAGCTACTAATAGGATAAATTATAGG
No	***Target Gene***	***Primer name***	***Primer sequence***
Primers for qPCR			
8	ALG9	ALG9-For1	GCCGTCTACGAGCAATTTTC
		ALG9-Rev1	ATCTGGCAGCAGGAAAGAAC
9	GFP	GFP-For1	TGGAAGCGTTCAACTAGCAG
		GFP-Rev1	AAAGGGCAGATTGTGTGGAC

PCR Purification Kit and measure the concentrations using NanoDrop.

5. Load the purified PCR products and the digestion product (1 μL DNA with 5 μL 6× loading dye) onto 1% agarose gel and perform electrophoresis at 120 V for 20 min. Image the gel with Molecular Imager Gel Doc system to verify the sizes of the corresponding DNAs.

6. The PCR products: Delta1, Delta2, and the GFP expression cassette 1, were assembled into plasmid pRS423 linearized with *Cla*I and *Eco*RI via Gibson Assembly to result in plasmid pSS99. The assembly reaction conditions were set according to the manual of Gibson Assembly Cloning Kit (*see* **Note 5**). Incubate the mixture at 50 °C for 1 h in a thermocycler. The assembled mixture was transformed into chemical competent *E. coli* DH5α as described above. Plate the transformed *E. coli* onto LB + Amp plates as described earlier. Incubate at 37 °C overnight.
7. Plasmid preparation. Inoculate three colonies into 5 mL LB + Amp medium and grow overnight at 37 °C, 250 rpm. Isolate the plasmids from the overnight culture using QIAprep Miniprep Kit and verify the plasmid by restriction digestion or sequencing.
8. Label the verified plasmids as pSS99. Plasmid pSS99 was linearized by *Avr*II digestion to generate donor DNA 8 kb (Fig. 1c). Digestion condition: 1 μL of *Avr*II, 5 μL of 10× CutSmart buffer, and 1 μg of pSS99 plasmid. Add ddH_2O to a final volume of 50 μL. Incubate at 37 °C for 3 h in a PCR thermocycler. Purify the digestion products using QIAquick PCR Purification Kit.
9. The BDO expression cassette was PCR amplified from plasmid pC77 [14] with primers: BDO-For1 and BDO-Rev1 (Table 1). The PCR reaction conditions were set according to the manual of KOD DNA Polymerase. Purify the PCR products using QIAquick PCR Purification Kit and measure the concentrations using NanoDrop.
10. Digest pSS99 plasmid at 37 °C for 6 h in a thermocycler. Reaction condition: 1 μL of *Sal*I, 1 μL of *Bsp120*I, 5 μL of 10× CutSmart buffer, and 1 μg of the pSS99 plasmid. Add ddH_2O to a final volume of 50 μL. Purify the digestion product using QIAquick PCR Purification Kit and measure the concentrations using NanoDrop.
11. Load the purified PCR product of BDO expression cassette and the digested pSS99 (1 μL DNA with 5 μL 6× loading dye) onto 1% agarose gel and perform electrophoresis at 120 V for 20 min. Image the gel with Molecular Imager Gel Doc system to verify the sizes of the corresponding DNAs.
12. The BDO expression cassette was then assembled into *Sal*I/*Bsp120*I digested plasmid pSS99 via Gibson Assembly to generate plasmid pSS100. The assembly reaction conditions were set up as described earlier and the assembled mixture was transformed as before and plated on LB + Amp plates.

13. Pick and grow three *E. coli* colonies in 5 mL of LB + Amp medium at 37 °C and 250 rpm overnight. Plasmids were extracted from overnight culture using QIAprep Miniprep Kit, and confirmed by enzyme digestion or sequencing.
14. Label the verified plasmids as pSS100. Plasmid pSS100 was linearized by *Avr*II digestion to generate donor DNA 16 kb (Fig. 1c). Digestion condition: 1 μL of *Avr*II, 5 μL of 10× CutSmart buffer, and 1 μg of pSS100 plasmid. Add ddH_2O to a final volume of 50 μL. Incubate at 37 °C for 3 h in a PCR thermocycler. Purify the digestion products using QIAquick PCR Purification Kit.
15. The xylose utilization cassette was PCR amplified from plasmid pRS426m-xylose-zeaxanthin with a pair of primers: Xylose-For1 and Xylose-Rev1 (Table 1) according to the manual of KOD DNA Polymerase. The PCR product was then purified with QIAquick PCR Purification Kit and the concentration was measured using NanoDrop.
16. Digest pSS100 plasmid at 37 °C for 6 h in a thermocycler. Reaction condition: 1 μL of *Sac*II, 1 μL of *BamHI*, 5 μL of 10× CutSmart buffer, and 1 μg of the pSS100 plasmid. Add ddH_2O to a final volume of 50 μL. Purify the digestion product using QIAquick PCR Purification Kit and measure the concentrations using NanoDrop.
17. Load the PCR product of xylose utilization cassette and the digestion pSS100 onto 1% agarose gel and perform electrophoresis at 120 V for 20 min. Image the gel with Molecular Imager Gel Doc system to verify the sizes of the corresponding DNAs.
18. The xylose utilization cassette was then assembled into *Sac*II/*BamHI* digested plasmid pSS100 via Gibson Assembly to generate pSS101. The assembly reaction conditions were set up as described earlier and the assembled mixture was transformed as before and plated on LB + Amp plates.
19. Pick and grow three *E. coli* colonies in 5 mL of LB + Amp medium at 37 °C and 250 rpm overnight. Plasmids were extracted from overnight culture using QIAprep Miniprep Kit, and confirmed by enzyme digestion or sequencing. Label the verified plasmids as pSS101. Plasmid pSS101 was linearized by *Avr*II digestion to generate donor DNA 24 kb (Fig. 1c). Digestion condition: 1 μL of *Avr*II, 5 μL of 10× CutSmart buffer, and 1 μg of pSS101 plasmid. Add ddH_2O to a final volume of 50 μL. Incubate at 37 °C for 3 h in a PCR thermocycler. Purify the digestion products using the QIAquick PCR Purification Kit.

3.3 Transformation

Transform pDi-CRISPR plasmid and linearized donor DNAs to strain HZ848 using the LiAC/SS carrier DNA/PEG method [18]. 300 ng of pDi-CRISPR was transformed together with equimolar amounts of linearized donor DNA [either 8 kb (300 ng), 16 kb (600 ng), or 24 kb (900 ng) constructs were used respectively] (*see* **Note 6**).

1. The night before the transformation, inoculate a colony of the *S. cerevisiae* HZ848 strain into 3 mL YPD medium and grow overnight at 30 °C and 250 rpm.
2. Add 1 mL of the overnight culture into 25 mL YPD medium in a baffled flask. Grow for 4–6 h at 30 °C and 250 rpm until OD_{600} (determined using the spectrophotometer) is approximately 1.
3. In the meantime, boil 500 μL of ssDNA for 5 min in a heating block and chill immediately on ice.
4. Prepare transformation mix containing 240 μL of 50% (w/v) PEG 3350, 36 μL of 1 M LiAc, and 50 μL of boiled carrier ssDNA per transformation reaction. Vortex vigorously on a vortex mixer. For more transformations, scale up transformation mix accordingly.
5. Prepare DNA mix containing 300 ng of pDi-CRISPR and either 300 ng 8 kb, 600 ng 16 kb or 900 ng 24 kb donor DNA. Make up the DNA mixtures to 40 μL using sterile ddH_2O.
6. When the OD_{600} of the cells have reached approximately 1, harvest the cells by centrifugation at 3000×*g* for 2 min in a benchtop centrifuge. Discard the YPD and resuspend the cell pellet in 10 mL of sterile ddH_2O by gentle pipetting/vortexing and centrifuge as before. Discard the water and resuspend pellet in 1 mL of LiAC/TE.
7. Transfer the cell suspension to a 1.5 mL Eppendorf microcentrifuge tube. Centrifuge at 12000 × *g* for 30 s in a microcentrifuge and discard the supernatant.
8. Add 350 μL of LiAc/TE and transfer 100 μL of the cell suspension into a fresh 1.5 mL microcentrifuge tube for each transformation reaction. Centrifuge the cell suspensions as before and discard the supernatant. Add 326 μL of the transformation mix into each tube and 40 μL of the appropriate DNA mix containing pCRCT and either 8 kb, 16 kb, or 24 kb DNA. Mix gently by pipetting.
9. Incubate the tube in 30 °C water bath for 30 min followed by 42 °C for 15 min (*see* **Note 7**).

10. Following incubation, centrifuge the cell suspensions as before in a microcentrifuge and discard the supernatant. Wash the cells once with water. Then load 2 mL of the appropriate SC media (SC-ura) to the washed cells and cultivate at 250 rpm, 30 °C. Once every 2 days, transfer 1 mL of cell culture to 1.2 mL of fresh media to ensure cell viability. On day 6, the cells were washed and prepared for flow cytometry analysis (*see* **Note 8**).
11. To estimate the transformation efficiency from colony counting, plate 10% of the freshly transformed cells on selective and nonselective plates. Grow cells for three days and count the numbers of fluorescent and nonfluorescent colonies with the help of the SZX16 fluorescent stereo microscope (green fluorescence mode) for determination of the percentages of GFP-fluorescent colonies. Obtain images of the colonies with the stereo microscope. Adjust the brightness of the green channel to account for autofluorescence from cells not containing GFP. Process all images using the same settings.

3.4 Flow Cytometry Analysis

1. Wash cultured cells by spinning down in a benchtop centrifuge at 3000 × *g* for 2 min. Remove cell culture medium and resuspend cells in 5 mL sterile PBS. Spin down cells again using the same settings, remove PBS and resuspend cells in 5 mL PBS. Check OD_{600} of cells using spectrophotometer and dilute cells to an OD_{600} of between 0.1 and 0.2.
2. Setting up the flow cytometer run. Use nonfluorescent HZ848 cells to set up the run. Briefly, start a new experiment in the BD FACSDIVA software. Under the "Parameters" tab, select "FCS-W," "SSC-W," and "GFP" as fluorophore. Create a new specimen and load nonfluorescent HZ848 into a tube holder before starting low-speed run on machine. Create SSC-A versus FSC-A dot plot (Fig. 3a). Draw a polygonal gate around cells and label the gated population as "cells". Select "cells" and create SSC-W vs FSC-W dot plot (Fig. 3b). Adjust voltage such as FSC-W and SSC-W falls between 50 and 100 (×1000) as shown in Fig. 3b. Draw rectangular gate around single cells and label the gated population as "singlets." Select "singlets" and create GFP-A histogram. Adjust excitation voltage (488 nm) so that GFP-A of nonfluorescent control falls close to 10^2 as shown in Fig. 3c. Draw gate to demarcate fluorescent cells and label the gated population as "GFP Pos." Set the "Events To Record" as 10,000. To save the plots, select "Acquire Data" followed by "Record Data" on the software.

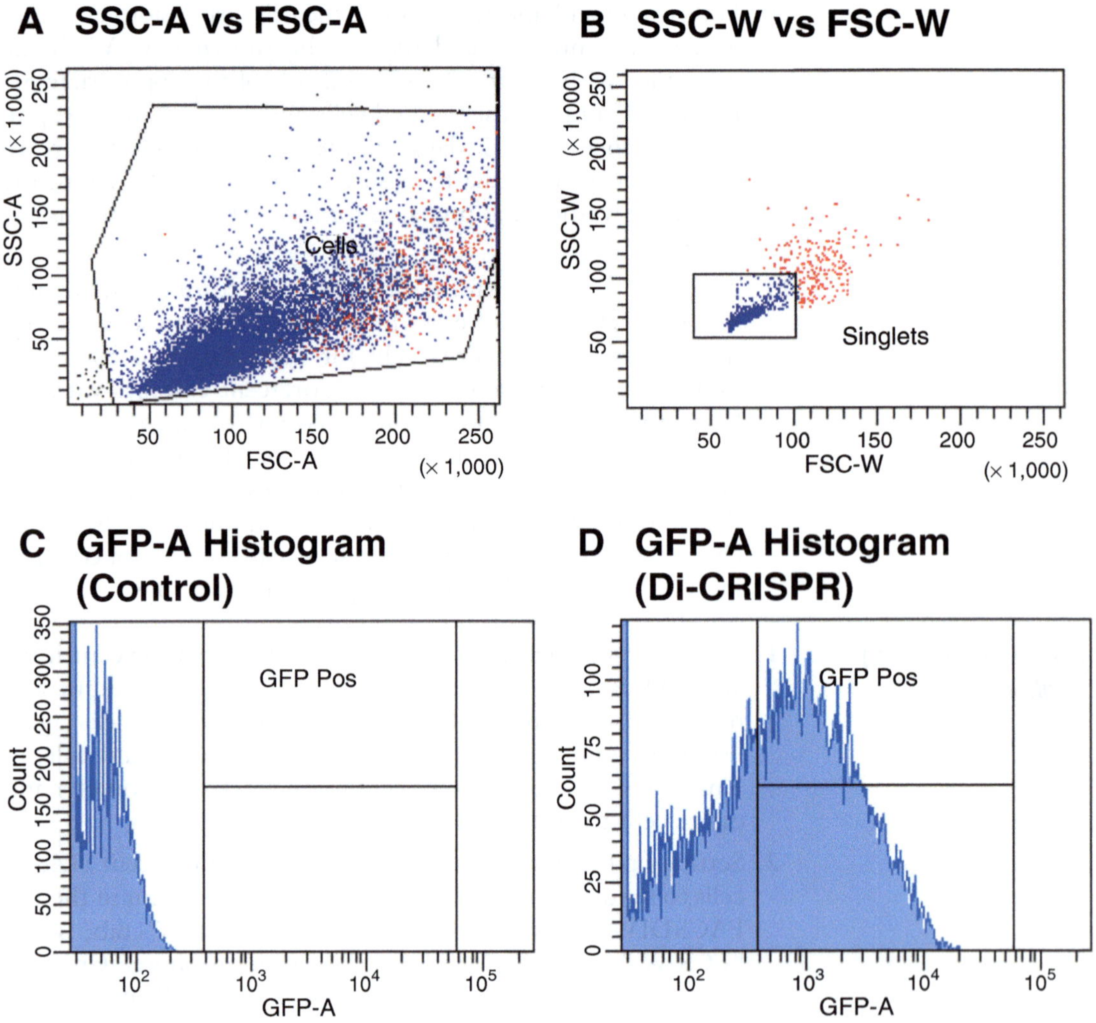

Fig. 3 Flow cytometry analysis. (**a**) SSC-A versus FSC-A dot plot. "Cells" are demarcated with polygonal gate. (**b**) SSC-W vs FSC-W dot plot. "Singlets" are demarcated with rectangular gate. (**c**) GFP-A histogram of nonfluorescent control. Nonfluorescent control was used to set "GFP Pos" gate for GFP-positive cells. (**d**) An example of GFP-A histogram of Di-CRISPR-treated cells. There is a larger proportion of GFP-positive cells following integration of donor DNA using our Di-CRISPR system

3. After analyzing the control, remove the tube from the tube holder and wash the setup with DI water. Load the next sample and repeat the acquisition by creating new specimen and selecting "Acquire Data" followed by "Record Data" to obtain GFP-A histogram (Fig. 3d) (*see* **Note 9**). Export the data in the desired format.

3.5 Quantitative PCR Analysis

1. Amplify GFP expression cassette 2 from plasmid pRS426-TEF1p-GEP-TEF1t with primers TEF1pFor2 and TEF1tRev2 (*see* **Note 10**). Integrate GFP expression cassette 2 at the *LYP1* site of *S. cerevisiae* HZ848 to generate a control strain for qPCR analysis using the LiAC/SS carrier DNA/PEG transformation method as described earlier [18]. The *LYP1* disruption mutant can grow in the presence of thialysine (250 μg/mL), and contain GFP expression cassette 2.
2. Inoculate the control strain and the newly constructed integration strains generated by Di-CRISPR into 3 mL YPD medium and grow overnight at 30 °C and 250 rpm. Isolate the genomic DNA from 3 mL cell culture using Wizard Genomic DNA Purification Kit. Measure the DNA concentrations using NanoDrop.
3. Determine gene copy numbers by quantitative PCR (qPCR) using the yeast total DNA extracts. Choose the *GFP* gene in the integration cassette and *ALG9* gene in the chromosome as the target and reference genes respectively. Adopt the control strain containing one copy of *GFP* and *ALG9* as the template for standard curves. Quantify the copy numbers by comparing the C_q values of the target and reference genes. The sizes of haploid *S. cerevisiae* genome is 12,495,682 bp. Determine genomic DNA mass using the following equation modified from Whelan et al. [19].

$$\text{DNA entity mass (g)} = \text{DNA entity size (bp)}\left[\frac{\text{dsDNA MW}_{\text{av}}\ (\text{g/mol})}{\text{Avogadro's No.}(\text{bp/mol})}\right]$$

where:DNA entity: genome or plasmid; dsDNA MW_{av}: average double-stranded DNA molecular weight = 617 g/mol; Avogadro's No. = 6.022×10^{23} bp/mol.

Therefore, the mass of genomic DNA of the newly constructed integration strain is 0.0128 pg, and it contains one copy of *ALG9* gene and unknown number of *GFP*; the mass of genomic DNA of the control strain is 0.0128 pg and it contains one copy of *ALG9* and *GFP*.

4. Perform qPCR using FastStart Essential DNA Green Master on a Roche Light Cycler® 96 System (*see* **Note 11**). Oligonucleotide primers for qPCR are listed in Table 1. They were designed and chosen from DNA fragments utilizing the program Primer 3 Version 0.4.0 [20]. Prepare qPCR reactions in triplicate in LightCycler® 480 Multiwell Plate 96. Each experimental reaction consists of 20 μL containing 10 μL of 2× FastStart Essential DNA Probes Master, 4 μL of 2 μM aliquots of each forward and reverse primer (for a final concentration of

400 nM), and 2 μL of a series of diluted genomic DNA as template. Determine DNA concentrations utilizing NanoDrop. Analyze a serial tenfold dilution, ranging from 60 to 6,000,000 copy numbers (*see* **Note 12**), of both unknown and control genomic DNA in triplicate to establish the standard curves. Set a negative control reaction for each pair of primers by substituting the template DNA with nuclease-free water. The PCR program consists of an initial denaturation at 95 °C for 10 min and then 45 cycles of two steps: 95 °C for 10 s and 60 °C for 20 s (*see* **Note 13**). Finally, keep the sample at 72 °C for 10 s. Use *ALG9* as reference, and calculate its absolute copy number from the amount of initial genomic DNA. The copy number of unknown *GFP* was calculated using the following quantification equation:

$$\frac{N_{0\ A_{tar}}}{N_{0\ A_{ref}}} = \frac{\left(1+E_{B_{ref}}\right)^{Cq_{Bref}}\left(1+E_{A_{tar}}\right)^{Cq_{Atar}}}{\left(1+E_{A_{ref}}\right)^{Cq_{Aref}}\left(1+E_{B_{tar}}\right)^{Cq_{Btar}}}$$

where: N_0: initial number of DNA molecules; A_{tar}: referred to a target gene (*GFP*) in the newly constructed integration strain; A_{ref}: referred to the reference gene (*ALG9*) in the newly constructed integration strain; B_{tar}: referred to a target gene (*GFP*) in control strain; B_{ref}: referred to the reference gene (*ALG9*) in the control strain; *E*: PCR efficiency; *Cq*: quantification cycle.

Obtain PCR efficiencies (*E*) from the slopes of standard curves (*see* **Note 14**). Each standard curve consists of a linearly fitted plot of threshold quantification cycles (C_q) vs. decimal logarithm of relative copy numbers. Determine C_q values after setting the threshold line of baseline subtracted fluorescence reading normalized to the reference dye in a semi-logarithmic scale for each standard curve.

4 Notes

1. Delta sequence: TGGAAGCTGAAACGTCTAACGGATCTT GATTTGTGTGGACTTCCTTAGAAGTAACCGAAGCACA GGCGCTACCATGAGAATTGGGTGAATGTTGAGATAATT GTTGGGATTCCATTGTTGATAAAGGCTATAATATTAGG TATACAGAATATACTAGAAGTTCTCCTCGAGGATATAGG AATCCTCAAAATGGAATCTGCAATTCTACACAATTCTAT AAATATTATTATCATCATTTTATATGTTTATATTCATTGA TCCTATTACATTATCAATCCTTGCGTTTCAGCTTCCAC TAATTTAGATGACTATTTCTCATCATTTGCGTCATCTT CTAACACCGTATATGATAATATACTAGTAATGTAAATAC TAGTTAGTAGATGATAGTTGATTTCTATTCCAACA.

2. The selected guide RNA (TATACTAGAAGTTCTCCTCG) cuts a total of 109 times in the reference *S. cerevisiae* (S288C) genome.
3. Alternatively, a double-stranded oligonucleotide can be ordered as a gBlocks Gene Fragment (IDT). The double-stranded DNA should be of the form GGTCTCCCCAAAA CNNNNNNNNNNNNNNNNNNNNGTTTTAGAGCGA GACC. Digest the gBlocks at 37 °C for 3 h in a 20 μL reaction containing 1 μL of *Bsa*I, 1 μL of 10× CutSmart Buffer, and 250 ng of gBlocks fragment. Purify the digestion products using QIAquick PCR Purification Kit into 12 μL of ddH_2O and measure the concentration using NanoDrop. Skip phosphorylation and annealing steps and proceed to ligate *Bsa*I-digested gBlocks fragment with *Bsa*I-digested pCRCT.
4. This rotation speed (250 rpm) can be used for most incubator shakers with orbit diameters ranging from 0.5 to 2 in. (1.27–5.1 cm).
5. To make successful cloning, researcher may need to optimize the assembly system. One of the most important parameters in a ligation reaction is the ratio of insert to vector molecules. For assembly of three or more fragments, it is recommended to use equimolar ratio of fragments. If plasmid template was used in the PCR reaction, it is recommended to treat the PCR product with *Dpn*I restriction endonuclease in order to digest dam methylated plasmid template before setting up the Gibson Assembly reaction.
6. For delta-integration, more donor DNA may be used to increase the integrated copy number if necessary.
7. Different *S. cerevisiae* strains may have different optimal heat-shock time.
8. The 6-day incubation period was chosen as it was found that fluorescence peaked and plateaued on that day presumably due to a delay in recovery and GFP expression.
9. If the GFP-A values of the fluorescent samples are so high that the GFP-A values fall off the histogram, lower excitation voltage and reanalyze all samples (including control) using the new settings.
10. Primers for amplification of the GFP expression cassette 2 contain homologous arms to *LYP1* site. The integration of GFP expression cassette 2 can support colony growth on plates with 250 μg/mL thialysine.
11. Keep the master mix away from light. Avoid repeated freezing and thawing.
12. Use up to 100 ng genomic DNA for a reaction of 20 μL. Using too much of DNA might reduce the fluorescence signal by outcompeting the dye.

13. The primer annealing temperature is primer dependent, and it is usually set 5 °C below the calculated primer Tm. For higher reaction volumes, it can be advantageous to choose longer annealing and extension times for the amplification cycles.
14. *E* values usually ranged between 85 and 98%.

Acknowledgments

We thank Agency for Science, Technology, and Research, Singapore for supporting various research projects in the Metabolic Engineering Research Laboratory (MERL) through the Visiting Investigator Programme to H.Z (1535j00137).

References

1. Hong KK, Nielsen J (2012) Metabolic engineering of *Saccharomyces cerevisiae*: a key cell factory platform for future biorefineries. Cell Mol Life Sci 69:2671–2690
2. Nevoigt E (2008) Progress in metabolic engineering of *Saccharomyces cerevisiae*. Microbiol Mol Biol Rev 72(3):379–412
3. Karim AS, Curran KA, Alper HS (2013) Characterization of plasmid burden and copy number in *Saccharomyces cerevisiae* for optimization of metabolic engineering applications. FEMS Yeast Res 13(1):107–116
4. Da Silva NA, Srikrishnan S (2012) Introduction and expression of genes for metabolic engineering applications in *Saccharomyces cerevisiae*. FEMS Yeast Res 12(2):197–214
5. Storici F, Durham CL, Gordenin DA, Resnick MA (2003) Chromosomal site-specific double-strand breaks are efficiently targeted for repair by oligonucleotides in yeast. Proc Natl Acad Sci U S A 100(25):14994–14999
6. de Jong B, Shi S, Valle-Rodríguez J, Siewers V, Nielsen J (2015) Metabolic pathway engineering for fatty acid ethyl ester production in *Saccharomyces cerevisiae* using stable chromosomal integration. J Ind Microbiol Biotechnol 42(3):477–486
7. Lee FWF, Silva NAD (1997) Improved efficiency and stability of multiple cloned gene insertions at the δ sequences of *Saccharomyces cerevisiae*. Appl Microbiol Biotechnol 48(3):339–345
8. Yamada R, Taniguchi N, Tanaka T, Ogino C, Fukuda H, Kondo A (2010) Cocktail δ-integration: a novel method to construct cellulolytic enzyme expression ratio-optimized yeast strains. Microb Cell Factories 9:32
9. Christian M, Cermak T, Doyle EL, Schmidt C, Zhang F, Hummel A et al (2010) Targeting DNA double-strand breaks with TAL effector nucleases. Genetics 186(2):757–761
10. Li T, Huang S, Zhao X, Wright DA, Carpenter S, Spalding MH et al (2011) Modularly assembled designer TAL effector nucleases for targeted gene knockout and gene replacement in eukaryotes. Nucleic Acids Res 39(14):6315–6325
11. Jakočiūnas T, Bonde I, Herrgård M, Harrison SJ, Kristensen M, Pedersen LE et al (2015) Multiplex metabolic pathway engineering using CRISPR/Cas9 in *Saccharomyces cerevisiae*. Metab Eng 28:213–222
12. DiCarlo JE, Norville JE, Mali P, Rios X, Aach J, Church GM (2013) Genome engineering in *Saccharomyces cerevisiae* using CRISPR-Cas systems. Nucleic Acids Res 41(7):4336–4343
13. Bao Z, Xiao H, Liang J, Zhang L, Xiong X, Sun N et al (2015) Homology-integrated CRISPR–Cas (HI-CRISPR) system for one-step multigene disruption in *Saccharomyces cerevisiae*. ACS Synth Biol 4(5):585–594
14. Shi S, Liang Y, Zhang MM, Ang EL, Zhao H (2016) A highly efficient single-step, markerless strategy for multi-copy chromosomal integration of large biochemical pathways in *Saccharomyces cerevisiae*. Metab Eng 33:19–27
15. Luo Y, Huang H, Liang J, Wang M, Lu L, Shao Z et al (2013) Activation and characterization of a cryptic polycyclic tetramate macrolactam biosynthetic gene cluster. Nat Commun 4:2894
16. Sun J, Shao Z, Zhao H, Nair N, Wen F, Xu J-H et al (2012) Cloning and characterization of a panel of constitutive promoters for

applications in pathway engineering in *Saccharomyces cerevisiae*. Biotechnol Bioeng 109 (8):2082–2092

17. Du J, Yuan Y, Si T, Lian J, Zhao H (2012) Customized optimization of metabolic pathways by combinatorial transcriptional engineering. Nucleic Acids Res 40(18):e142–e142
18. Gietz RD, Schiestl RH (2007) High-efficiency yeast transformation using the LiAc/SS carrier DNA/PEG method. Nat Protoc 2:31–34
19. Whelan JA, Russell NB, Whelan MA (2003) A method for the absolute quantification of cDNA using real-time PCR. J Immunol Methods 278(1):261–269
20. Rozen S, Skaletsky H (2000) Primer3 on the WWW for general users and for biologist programmers. In: Bioinformatics methods and protocols: methods in molecular biology. Edited by Krawetz S, Misener S. Humana Press, Totowa, NJ, pp 365–386

Chapter 7

Assembly of Complex Pathways Using Type IIs Restriction Enzymes

Sylvestre Marillonnet and Stefan Werner

Abstract

Efficient DNA assembly methods are essential tools for synthetic biology and metabolic engineering. Among several recently developed methods that allow assembly of multiple DNA fragments in a single step, DNA assembly using type IIS enzymes provides many advantages for complex pathway engineering. In particular, it provides the ability for the user to quickly assemble multigene constructs using a series of simple one-pot assembly steps starting from libraries of cloned and sequenced parts. We describe here a protocol for assembly of multigene constructs using the modular cloning system (MoClo). Making constructs using the MoClo system requires to first define the structure of the final construct to identify all basic parts and vectors required for the construction strategy. Basic parts that are not yet available need to be made. Multigene constructs are then assembled using a series of one-pot assembly steps with the set of identified parts and vectors.

Key words Modular cloning, Synthetic biology, Biological parts, DNA assembly, Multigene constructs

1 Introduction

For many years, constructs were made using protocols that required performing a series of steps that included DNA digestion, separation of DNA fragments on a gel, extraction of DNA fragments from the gel, ligation of the fragments, transformation of the ligation in *E. coli*, and screening of the colonies obtained to find one that contained the expected construct. This process used to be laborious and time consuming. Fortunately, it can now be replaced by more efficient strategies that include homology-based or type IIS-based assembly methods, the latter of which are being presented here in this chapter. Type IIS-based methods rely on Golden Gate cloning, a method that allows assembly of multiple DNA fragments using a simple one-pot one-step assembly reaction. Golden Gate cloning combines the use of a type IIS enzyme with a restriction–ligation [1]. Type IIS enzymes, such as BsaI or BpiI,

Christine Nicole S. Santos and Parayil Kumaran Ajikumar (eds.), *Microbial Metabolic Engineering: Methods and Protocols*, Methods in Molecular Biology, vol. 1927, https://doi.org/10.1007/978-1-4939-9142-6_7,

are characterized by the fact that restriction enzyme cleavage site lies outside of the enzyme DNA recognition site. This allows DNA digestion and ligation to be performed simultaneously if the donor fragment and the recipient vector are designed in such a way that all sites for the type IIS enzyme used for cloning are eliminated during the cloning reaction. Golden Gate has been used as the base of several related but not identical modular cloning systems for plants [2–8], yeast and fungi [9–13] and bacteria [14, 15]. These cloning systems may vary by the specific type of IIS enzymes used for cloning in the different assembly steps, by the sequences of the type IIS cleavage sites that are used for cleavage and assembly, or by the antibiotic selection markers present on the various vector backbones. We present here a protocol for the modular system MoClo that was first designed to be used in plants [2], but can easily be adapted for use in other organisms such as yeast and bacteria. The MoClo system is a hierarchical assembly system by which transcription units are assembled from a set of level 0 modules that contain basic genetic elements such as promoters, coding sequences and terminators. The resulting level 1 constructs are then assembled in multigene constructs using two alternative strategies: (1) assembly in level 2 constructs by successive addition of sets of up to six level 1 transcription units to a growing construct, or (2) assembly of several intermediate level M multigene constructs, which are then assembled in a level P construct (Fig. 1) (*see* **Note 1**). As we have already described in detail a protocol for assembly of level 2 constructs [16], this chapter is designed to describe how to make multigene constructs using level M and P constructs.

2 Materials

2.1 Cloning

1. Restriction endonuclease *Bsa*I (10 U/μL) (NEB, New England Biolabs Inc., Ipswich, MA, USA) (*see* **Note 2**).
2. Restriction endonuclease *Bpi*I (10 U/μL) (ThermoFisher Scientific, St. Leon-Rot, Germany).
3. Restriction endonuclease *BsmB*I (10 U/μL) (NEB).
4. T4 DNA Ligase 3 U/μL or T4 DNA Ligase (HC) 20 U/μL, both supplied with 10× ligation buffer (300 mM Tris–HCl pH 7.8, 100 mM $MgCl_2$, 100 mM DTT, 10 mM ATP).
5. For measuring of DNA concentration, we use the NanoDrop ND2000 (Peqlab, Erlangen).
6. Luria–Bertani (LB) medium: 1% (*w/v*) Bacto tryptone, 0.5% (*w/v*) yeast extract, 1% (*w/v*) NaCl in deionized water, adjusted to pH 7.0 with 5 N NaOH. For plates, 1.5% (*w/v*) agar is added.

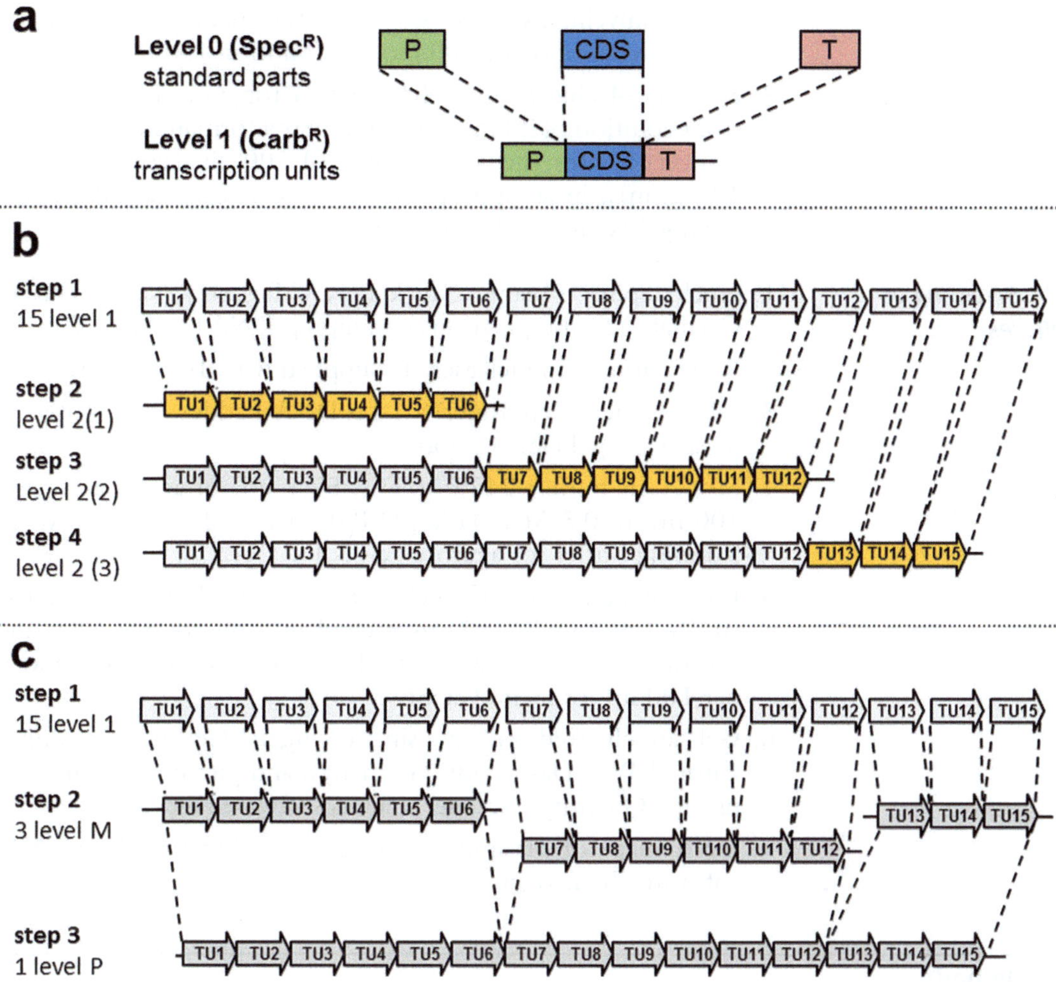

Fig. 1 Comparison of multigene assembly using level 2 versus level M and P. (**a**) Level 1 constructs are assembled from standard parts (level 0 modules) such as promoters (P), coding sequences (CDS), and terminators (T) using a first Golden Gate cloning reaction, resulting in constructs containing a transcription unit (TU). To make a construct containing 15 TUs (as an example), 15 level 1 constructs (represented as arrows in **b** and **c**) are assembled in parallel from selected level 0 modules. (**b**) Assembly of multigene constructs using level 2 constructs. Up to six level 1 constructs are assembled at each step into a level 2 construct that grows larger at each step. For example, steps 2–4 show the successive addition of 6, 6, and 3 transcription units to a level 2 construct (transcription units cloned directly from level 1 constructs are shown in yellow). (**c**) Assembly of multigene constructs using level M and P constructs. Groups of up to six transcription units are preassembled in level M constructs. The level M constructs are then assembled in a level P construct

7. Antibiotics carbenicillin (used instead of ampicillin) and kanamycin: filter-sterilized stocks of 50 mg/mL in H_2O (stored in aliquots at −20 °C) are diluted 1:1000 (final concentration: 50 μg/mL) in an appropriate amount of medium after the medium has been autoclaved and cooled down. For

spectinomycin, a stock of 40 mg/mL is made and is used at a final concentration of 100 μg/mL (dilution 1:400).

8. 5-bromo-4-chloro-3-indolyl-β-D-galactopyranoside (X-gal): stock solution of 20 mg/mL in dimethylformamide. For preparation of plates, the stock is diluted 1:500 (final concentration: 40 μg/mL) in an appropriate amount of LB agar after autoclaving/melting and cooling down.

2.2 Screening of Colonies

1. NucleoSpin® Plasmid Quick Pure (Macherey Nagel, Düren, Germany) for preparation of miniprep DNA (*see* **Note 3**).
2. Restriction endonucleases, all supplied with 10× buffer.
3. DNA ladder: GeneRuler™ 1 kb DNA Ladder Plus is used as marker for gel electrophoresis.
4. 50× TAE buffer: 242.0 g Tris, 57.1 mL of acetic acid, and 100 mL of 0.5 M EDTA, pH 8.0, in 1 L of deionized water. Running buffer for agarose gels is 1× TAE.
5. For preparation of gels for electrophoresis, agarose (0.7%–1.5%) in 1× TAE is melted in a microwave oven and one drop of 0.025% (w/v) ethidium bromide solution is added per 100 mL of melted agarose.
6. Gels are checked visually using a Syngene GelVue transilluminator (VWR, Darmstadt, Germany), and pictures are taken by using a Quantity one® gel analysis software (Bio-Rad).
7. DNA maps of plasmids are made by using the Vector NTI software (Invitrogen).

3 Methods

The first and perhaps most important step in the assembly of multigene constructs is planning the structure of the final construct and determining the assembly strategy. For planning the construct, one needs to consider the composition of each transcription unit (or gene), the number of genes, the order and orientation of the genes in the assembled construct, and the potential use of a selection marker for stable plant transformation (which, in fact, simply consists of adding one more gene to the multigene construct). There is no theoretical limit to the number of genes that can be assembled; however, the limit is set by the size and the stability of the plasmid which has to be cloned and maintained in *E. coli*. The process of selecting standard or individualized parts for the transcription units, the choice of destination vectors, and their assembly into higher order constructs are described below.

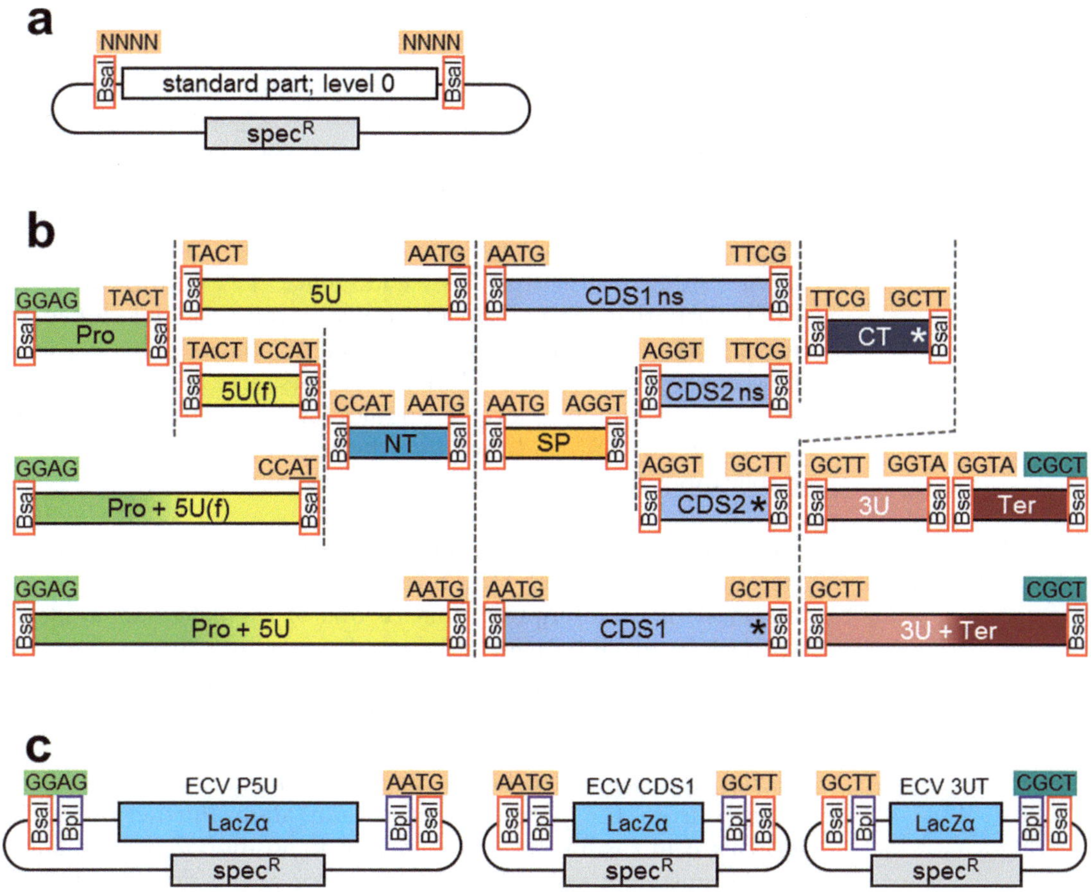

Fig. 2 Level 0 modules (standard parts) and cloning vectors. (**a**) General structure of a level 0 plasmid. The level 0 module is flanked by two inverted *BsaI* sites. (**b**) Different types of standard parts with their corresponding fusion sites. Promoter (Pro), 5′- and 3′-UTR (5 U, 3 U), N- and C-terminal tag (NT, CT), signal peptide (SP), coding sequence (CDS) with stop codon (*) or without (ns), terminator (Ter). (**c**) Level 0 cloning vectors. Only three vectors are shown as an example, but vectors for each module type are available

3.1 Selection of Standard Parts and Vectors

3.1.1 Selection of Standard Parts

Assembly of multigene constructs starts with the assembly of transcription units (genes) from basic parts (Fig. 1). MoClo basic parts, also called level 0 modules or standard parts, consist of a DNA fragment of interest flanked by two *Bsa*I restriction sites in opposite orientations cloned in a conventional cloning vector with a spectinomycin resistance cassette (Fig. 2a). The various types of level 0 modules comprise, but are not limited to, promoters, 5′-untranslated regions (UTRs), coding sequences, 3′-UTRs and terminators (*see* **Note 4**). The 4-bp sequences of the *Bsa*I cleavage sites (the "fusion sites") are specific for each type of module and define which parts can be fused together. For example, a promoter module (with TACT as 3′ fusion site) can be fused only to an untranslated leader sequence with the same fusion site at the 5′

end. The different types of standard parts and their fusion sites are shown in Fig. 2b.

Making a multigene construct first requires identifying all basic parts needed. Several parts may already be available; for example, the promoters or terminators may have already been cloned by the same or another laboratory. For example, a selection of 96 standard biological parts (including plant promoters and terminators, tags, reporter genes, and selectable markers) can be obtained from Addgene (https://www.addgene.org/cloning/MoClo/) [17]. Parts that are not already available can easily be cloned by users into corresponding level 0 cloning vectors (Fig. 2c; level 0 vectors and all other MoClo vectors are also available at Addgene). Cloning of level 0 modules has been described in detail earlier [2, 17, 18], and therefore the reader should refer to these publications for a description of that process (*see* **Note 5**).

3.1.2 Selection of Level 1 Destination Vectors

The position and orientation of each gene in a final construct determine which level 1 destination vector has to be chosen for assembly of a transcription unit. A total of 14 vectors are available (pL1F1-7, pL1R1-7) for cloning of genes in either forward or reverse orientation at each of seven possible positions (Fig. 3). These 14 level 1 vectors differ only by the sequence of the fusion sites. The two external fusion sites (*Bpi*I cleavage sites) of each vector are designed to be compatible with the fusion sites of the vectors from the positions before and after. Furthermore, the downstream fusion site of the vector at position 7 is compatible to the upstream fusion site of the vector at position 1 (TGCC). Thus, multigene constructs with less than seven genes can be assembled without the need for additional cloning vectors. In practice, a gene that is planned to be cloned at position 8 in a multigene construct is simply cloned in position 1 of the destination vector. Similarly, a gene at position 9 has to be cloned in position 2 of the destination vector and so on. An example for making a multigene construct containing 18 genes is shown in Fig. 4. All level 1 vectors have identical internal fusion sites (*Bsa*I cleavage sites) to allow assembly of the same level 0 parts in all vectors, that is, the first level 0 module must have GGAG as upstream fusion site (as with all promoter modules) and the last module must have CGCT as downstream fusion site (as with all terminator modules). An example for cloning a gene with three basic parts is shown in Fig. 5. All level 1 vectors are binary plasmids that can be used for *Agrobacterium*-mediated transient expression in plants. These vectors are nevertheless suitable for making constructs for other organisms, as level 1 constructs are usually used only as intermediate constructs, and the only feature that is required for this purpose is replication and selection in *E. coli*. Other vectors (for example, nonbinary vectors) could also be used

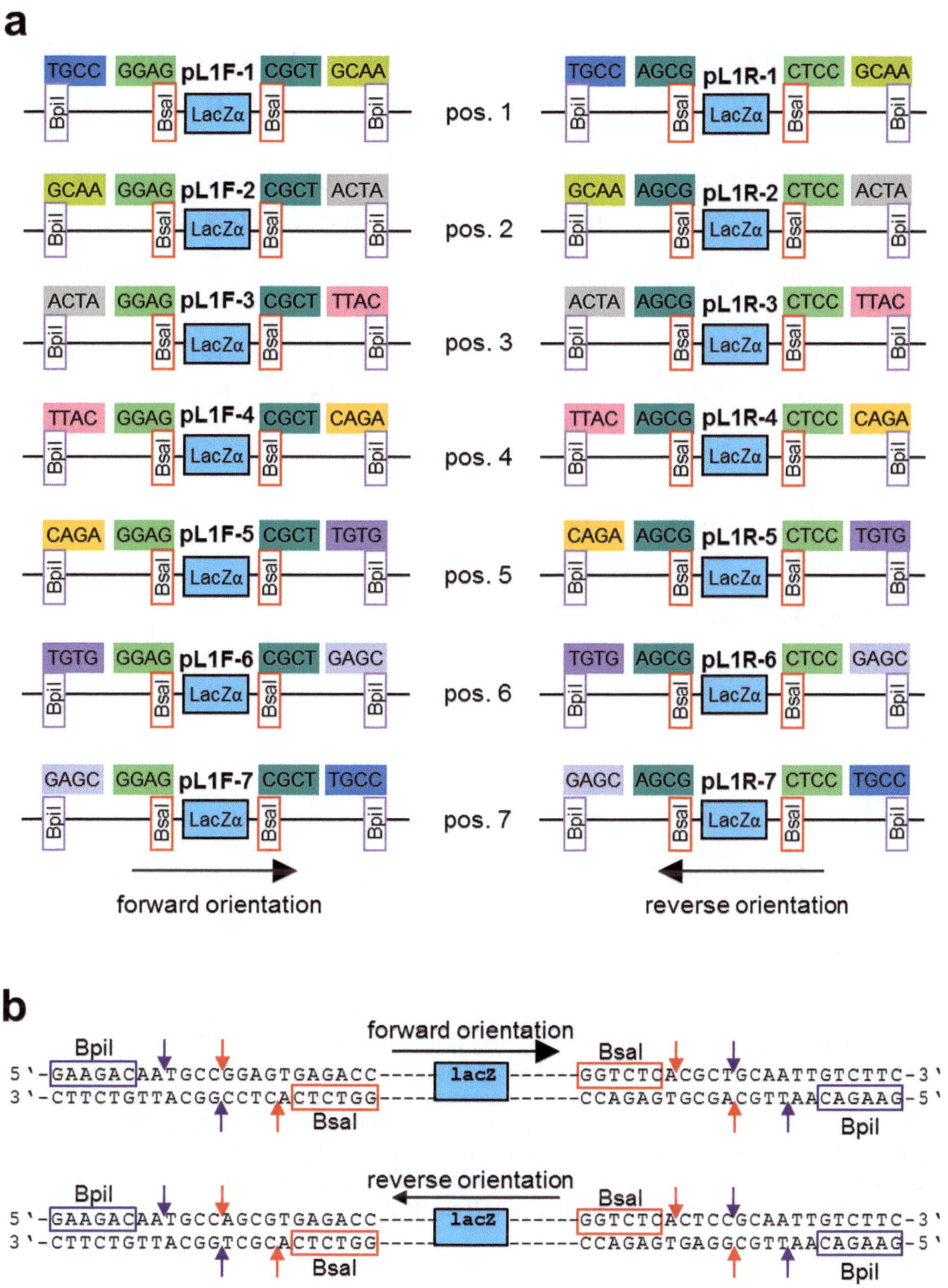

Fig. 3 Level 1 destination vectors. (**a**) Genes can be cloned at one of seven positions in forward (pL1F-vectors) or reverse (pL1R-vectors) orientation. (**b**) Sequence of the type IIS recognition (rectangles) and cleavage sites (arrows) of vectors pL1F-1 and pL1R-1

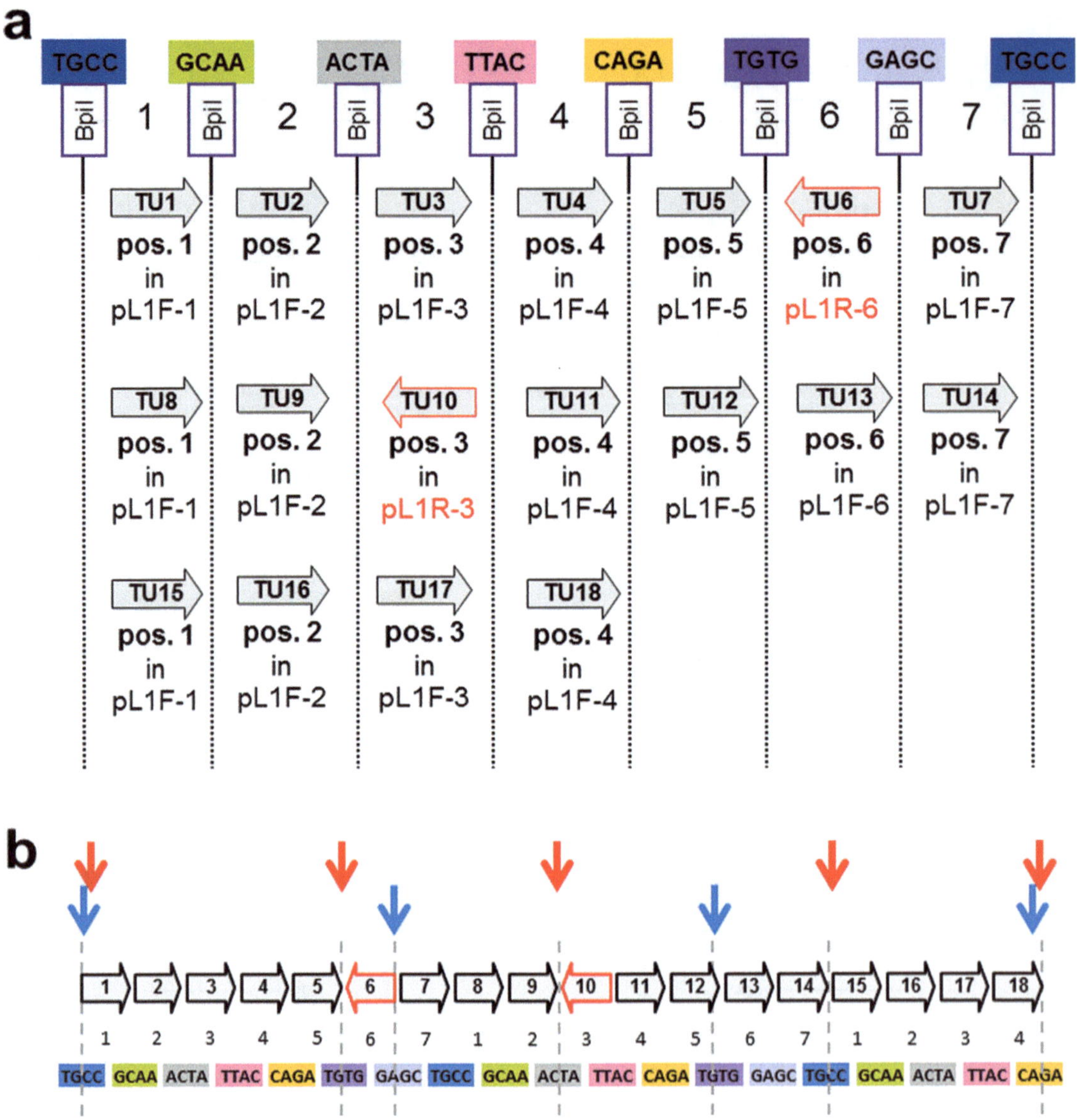

Fig. 4 Example for planning a multigene construct with 18 genes. (**a**) Assignment of transcription units (TU) to one of seven positions and the corresponding cloning vectors. Two genes (TU6, TU10) are cloned in reverse orientation. (**b**) Visual aid for planning level M and P constructs. The final construct is represented with the fusion sites flanking all level 1 constructs shown below the final construct. The same fusion sites will also flank the level M and P constructs. Subcloning groups of six level 1 constructs in three level M constructs (groups delimited by blue arrows) is valid as none of the fusion sites will be used twice to make the final level P construct. Subcloning of four level M constructs as shown with red arrows is not valid as fusion site TGCC would have to be used twice to make the final level P construct

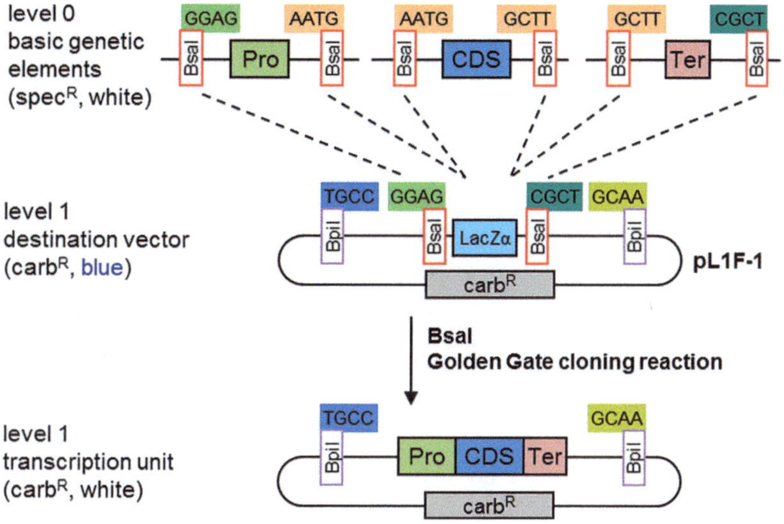

Fig. 5 Example of cloning a transcription unit consisting of three basic parts. Correct clones are identified by the antibiotic resistance marker (carb^R) and the color of the colonies on plates with X-gal (white)

as long as they contain the features outlined in Fig. 3 and an appropriate selectable marker (an ampicillin/carbenicillin resistance gene).

3.2 Assembly of Transcription Units

Once the level 0 modules and the destination vector are made or selected from previously made constructs, assembling the gene only requires pipetting all components into a reaction mix and incubating the reaction vessel in a thermal cycler.

1. Set up a restriction–ligation by pipetting into a tube 20 fmol (approximately 50 ng; *see* **Note 6**) of each level 0 module (*see* **Note 7**) and of the vector, 2 μL 10× ligation buffer, 10 U (1 μL) of *Bsa*I, and either 3 U (1 μL) of ligase for assembly of two to four modules or 20 U (1 μL) HC ligase for assembly of more than four modules (final volume of 20 μL; *see* **Note 8**).
2. Incubate the restriction–ligation mix in a thermal cycler. For assembly of two to four level 0 modules, incubation for 60–120 min at 37 °C is sufficient. If more modules are ligated together, the incubation time is increased to 6 h, or cycling is used as following: 2 min at 37 °C followed by 3 min at 16 °C, both repeated 50 times (*see* **Note 9**).
3. Restriction–ligation is followed by a digestion step (5 min at 50 °C) and then by heat inactivation for 10 min at 80 °C. The final incubation step at 80 °C is very important and is needed to inactivate the ligase at the end of the restriction–ligation. Omitting this step would lead to religation of some of the insert and plasmid backbone fragments when the reaction vessel is taken

out of the thermal cycler, and would lead to a higher proportion of colonies containing incorrect constructs.

4. Transform the entire ligation into chemically competent DH10B cells (*see* **Note 10**).
5. Thaw frozen chemically competent cells (100 μL per tube) on ice.
6. Add the entire ligation to the cells, and incubate on ice for 30 min.
7. Incubate 90 s at 42 °C in a water bath.
8. Let the cells recover on ice for 5 min.
9. Add 1 mL of LB medium to the cells, and incubate the tube at 37 °C in a shaker-incubator (150 rpm) for 45 min to 1 h.
10. After incubation, plate 25–100 μL of the transformation on LB agar plates containing carbenicillin and X-gal.
11. Incubate the plates overnight at 37 °C. Many white and very few blue colonies should be obtained.
12. Pick a few white colonies from the plate and transfer to liquid LB medium containing carbenicillin (*see* **Notes 11** and **12**).
13. Grow bacterial culture overnight at 37 °C and use for isolation of plasmid DNA.
14. Check plasmids by restriction digest using *Bpi*I (for MoClo level 1 modules) or any other suitable enzyme.

3.3 Assembly of Level M Constructs

Level 1 constructs are preassembled in groups of up to 6 into level M vectors. The resulting level M constructs will later be assembled in a Level P construct. Level M vectors contain two *Bpi*I recognition sites in opposite orientation for insertion of level 1 modules. The upstream fusion site defined by the first *Bpi*I site (for example, TGCC for pLM-1; Fig. 6) is compatible with fusion sites flanking upstream fusion sites of the level 1 vectors, whereas the downstream fusion site consists of a universal sequence (GGGA). This design allows cloning of two to six genes in the same vector (using more genes in this step would lead to incorrect clones because the same fusion site would be present on different modules). The last gene is fused to the vector using a compatible linker. We use here an example for the cloning of four transcription units in a final level P vector, with cloning of transcription units by groups of two in level M constructs (this small number of genes could be cloned directly in a final level M construct, but is chosen here for the sake of space in the figures). Cloning of two transcription units in the first level M construct will require using vector pLM1 and end-linker pELM-2 (*see* compatibility of the fusion sites in Figs. 6 and 7). Cloning of the next two transcription units will require vector pLM3 and end-linker pELM-4.

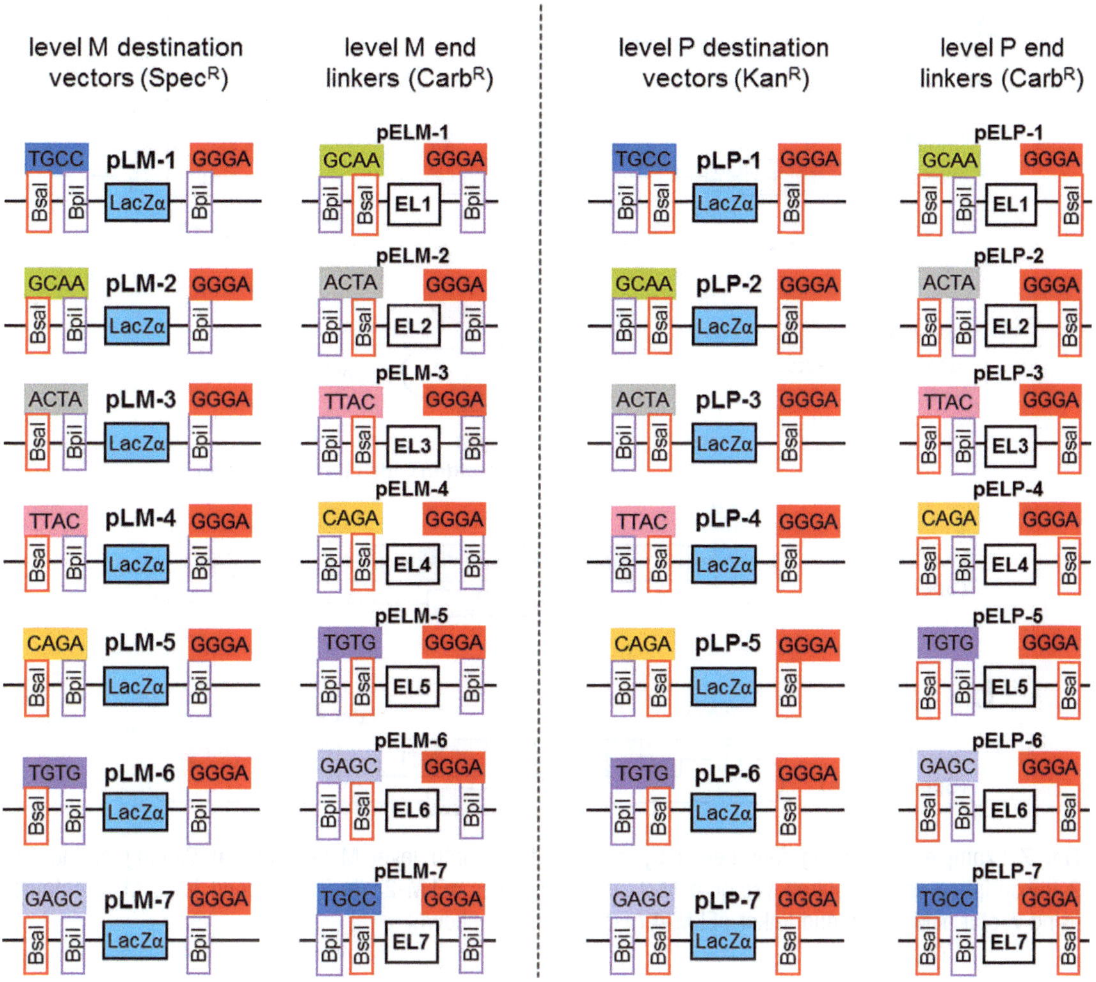

Fig. 6 Level M and P destination vectors and end-linkers

After cloning, level M constructs become flanked by two BsaI sites, one coming from the level M vector, and the other coming from the level M end-linker (Figs. 6 and 7). The fusion sites flanking these level M vectors are derived from those used previously in level 1 constructs. It is important to check, before starting cloning of level M constructs, that all level M constructs that are made will be compatible for assembly in level P vectors. We can take an example for defining a strategy for making a construct containing 18 genes (*see* Fig. 4b). One possibility is to clone the transcription units in level M vectors in three groups of 6 genes. This strategy is valid as the fusion sites that will flank the level M constructs (TGCC, GAGC, TGTG, and CAGA) are all different (Fig. 4b, blue arrows). Alternatively, one could decide to clone the level 1 constructs in groups of no more than five transcription units for higher cloning efficiency, for example in four groups of 5, 4, 5, and

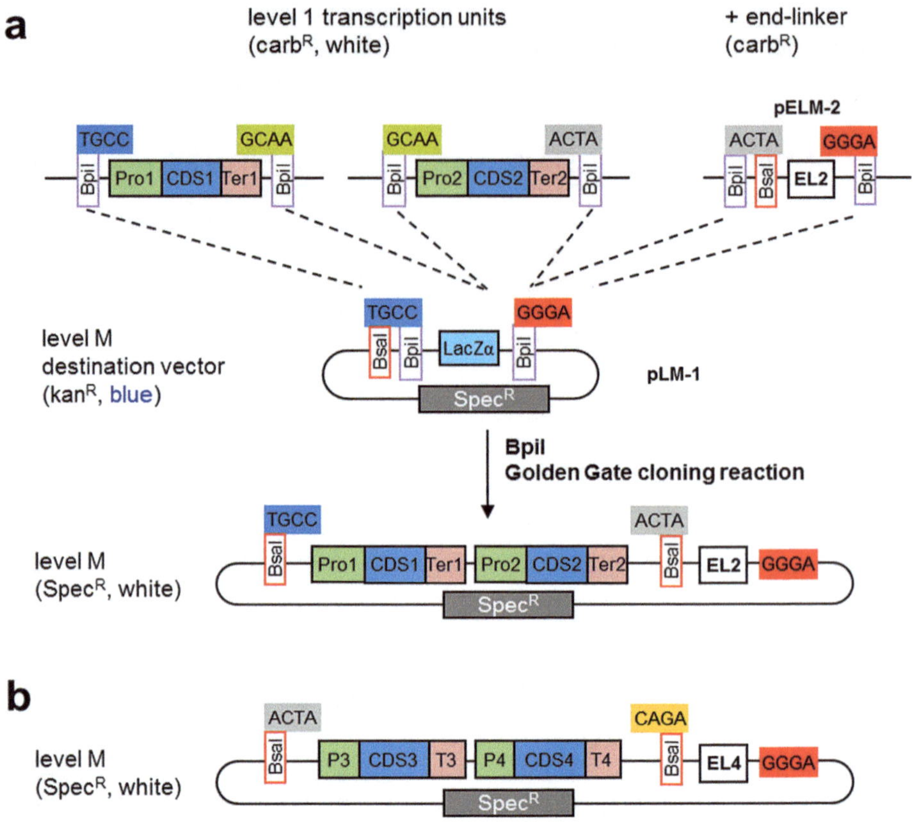

Fig. 7 Example for cloning four genes by groups of two into level M vectors. (**a**) Cloning of the first 2 transcription units requires vector pLM-1 and end linker pELM-2. (**b**) The second level M is obtained using vector pLM-3 and end linker pELM-4 (part details not shown)

4 transcription units. However, this strategy is not valid as the fusion sites that would flank the resulting level M constructs are TGCC, TGTG, ACTA, TGCC, and CAGA, with TGCC being present twice (Fig. 4b, red arrows). An alternative would be to clone groups of 5, 4, 4, and 5 transcription units, with fusion sites TGCC, TGTG, ACTA, GAGC, and CAGA flanking the resulting level M constructs.

The Golden Gate reaction for making level M constructs is virtually the same as described in Subheading 3.2 except that the enzyme *Bpi*I is used instead of *Bsa*I.

1. Set up a restriction–ligation by pipetting into a tube 20 fmol of each level 1 module, of the end-linker and of the level M vector, 2 μL 10× ligation buffer, 10 U (1 μL) of *Bpi*I, and either 3 U (1 μL) of ligase for assembly of two to four modules or 20 U (1 μL) HC ligase for assembly of more than four modules (final volume of 20 μL).
2. Incubate the restriction–ligation mix in a thermal cycler. For assembly of two to four level 1 modules, incubation for

60–120 min at 37 °C is sufficient. If more modules are ligated together, the incubation time is increased to 6 h, or cycling is used as following: 2 min at 37 °C followed by 3 min at 16 °C, both repeated 50 times (*see* **Note 9**).

3. Restriction–ligation is followed by a digestion step (5 min at 37 °C; *see* **Note 13**) and then by heat inactivation for 10 min at 80 °C.
4. Transformation of the reaction mix into *E. coli* is performed as described in Subheading 3.2, and cells are spread on LB plates with spectinomycin and X-gal. Colonies containing correct constructs should be white whereas colonies containing uncut or religated vector should be blue. Plasmids should be isolated from a few colonies and should be analyzed by restriction digest with a suitable enzyme.

3.4 Assembly of Level P Constructs

Level P vectors and end-linkers are similar to level M, but all BpiI sites and BsaI sites are reciprocally exchanged. In addition, level P vector have a different antibiotic selectable marker: Kan^R instead of $spec^R$ (Fig. 6). As for level M constructs, cloning in level P vectors requires selecting a suitable level P vector and an end-linker. In the example where 2 level M constructs consisting of two transcription units are cloned, vector pLP-1 and end-linker pELP-4 are used (Fig. 8). The Golden Gate reaction is the same as described in Subheading 3.2. Level P constructs are flanked by BpiI restriction sites and have therefore a similar structure as level 1 constructs (except for the selection marker). As level 1 constructs, several level P constructs can be assembled into level M vectors. As discussed with level M constructs, the user has to make sure that all level P vectors that need to be assembled are compatible with respect to their fusion sites. The suitability of the fusion sites has to be determined at the beginning of the project, before starting the process of making level M constructs.

4 Notes

1. Several level P constructs can be assembled to make larger constructs using level M vectors. This process can be repeated indefinitely by alternating level M and level P constructs. The limitation will of course be the size of the final construct and its ability to replicate in *E. coli*.
2. Type IIS restriction enzymes and ligase can be purchased from several commercial suppliers. Users should nevertheless be aware that various isoschizomers of type IIS restriction enzymes may not necessarily all cleave DNA with the same efficiency when using the reaction conditions described in this protocol (especially the restriction–ligation buffer). We have

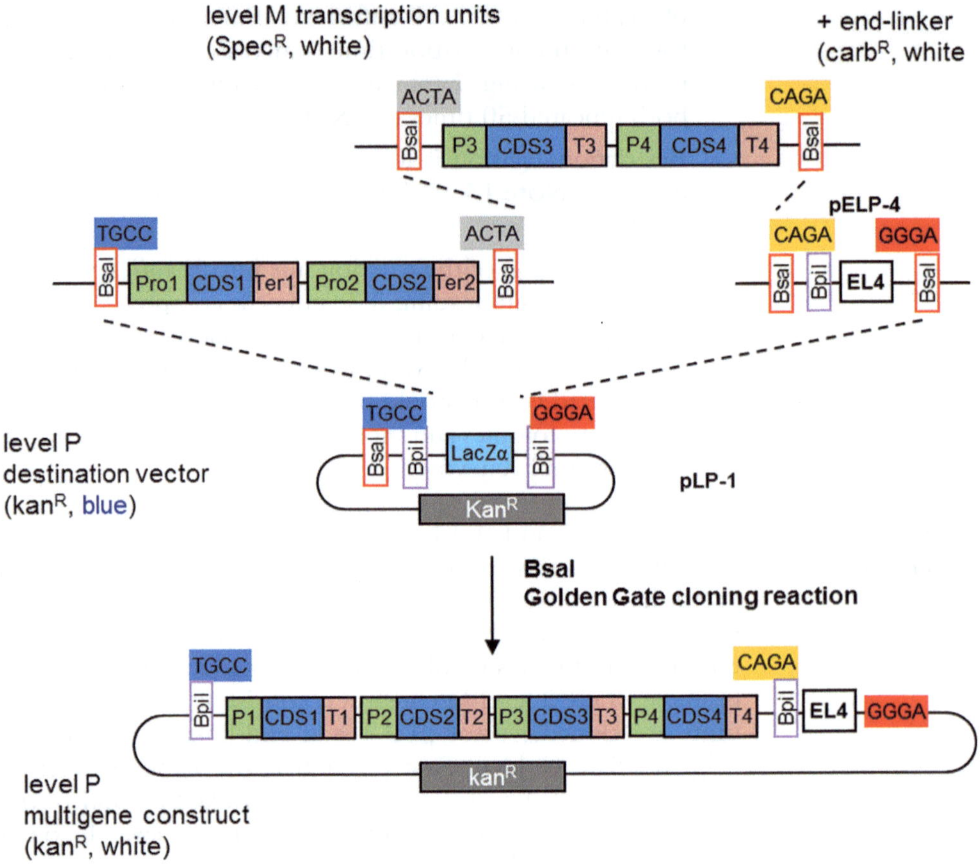

Fig. 8 Example of level P cloning from two level M constructs

obtained good results with the enzymes and reaction conditions described here, but use of enzymes from other sources might require adaptation of the reaction conditions to obtain optimal results. For example, with the conditions that we are using, better cloning efficiency is obtained when using BsaI rather than Eco31I.

3. Plasmid purification can be done with any commercially available kit that provides high purity DNA. Cloning of multigene constructs can result in very large plasmids for which standard miniprep kits are not suitable. In these cases, the use of kits designed for purification of large plasmids is recommended (e.g., NucleoBond® PC from Macherey-Nagel).
4. Some basic parts may be made for constructing something else other than a transcription unit. For example, a recombination site sequence may be cloned as a level 0 part between fusion sites GGAG and CGCT. In that case, the level 1 construct will be assembled from this part alone plus the appropriate level 1 vector.

5. An important point to consider is that level 0 modules should not contain recognition sites for the type IIS restriction enzymes used during the different cloning steps (BsaI, BpiI, and optionally BsmBI for use of the modules with level 2 vectors [16] or for other applications). If recognition sites for these enzymes are present, they will have to be removed using a process called domestication. Domestication is usually done by PCR amplification of the sequences of interest using primers designed to introduce point mutations (silent mutations for sites in coding sequences) in the restriction site sequences. Another possibility is to have the sequence lacking these restriction sites synthesized by a commercial gene synthesis provider. This solution provides the added benefit of allowing codon optimization (in the case of coding sequences), which might be useful for expression in nonnative host species.
6. In practice, if all module plasmids and the vector have approximately the same size (4–5 kb), simply adding 50 ng of DNA of each module and of the vector will work relatively well. However, when plasmids with widely different sizes are used, calculating an equimolar amount should provide a higher cloning efficiency. We have previously recommended using 40 fmoles of each fragment and of the vector for cloning. It however appears that using less DNA leads to higher cloning efficiency. We therefore now recommend using 20 fmoles for each insert and vector. The following formula (from the NEB catalog) can be used: 1 μg of a 1000 bp DNA fragment corresponds to 1.52 pmol. Therefore, the volume of DNA to pipet (in μL) to have 20 fmol is given by the equation: 20 (fmol) × size (bp) of the DNA fragment/(concentration (ng/μL) × 1520). Frequently, this formula will lead to volumes <1 μL with standard miniprep DNA. In these cases, we recommend diluting the DNA in order to avoid pipetting errors.
7. The Golden Gate cloning reaction is quite efficient. This allows making construct libraries rather than individual constructs. Construct libraries are made by using several modules of the same type (flanked by the same fusion sites) in the ligation mix rather than a single module of each type. In that case, if *n* variant modules replace a single module, the amount of each variant is divided by *n*, but the total amount of fragments for this module type remains the same, that is, 20 fmol. Constructs libraries can be made in vectors containing a ccdB marker for counter-selection to avoid growing empty vectors, but can also be made in normal constructs containing a standard LacZ marker if the presence of a few empty plasmid vectors in the resulting library is acceptable for the project.

8. The total volume of enzymes in the reaction should not exceed 10% (*v*/*v*). Otherwise, glycerol from the storage buffer of the enzymes can lead to decreased efficiencies.
9. Both the temperature conditions and ligase concentration are complementary parameters. For simple constructs requiring ligation of only one or two inserts, any program should work efficiently, even with ligase at regular concentration, and in most cases continuous incubation at 37 °C should be sufficient. The cycling program may improve the cloning efficiency especially when more modules are used and when plasmids become larger. For some applications that require high cloning efficiencies (e.g., the cloning of promoter libraries), use of high-concentration ligase and cycling are recommended.
10. Any other *E. coli* strain can also be used as long as it provides complementation for the lacZα fragment to allow blue-white selection. However, different *E. coli* strains show different transformation efficiencies which can be important for certain applications or when transforming large plasmids. In some cases and with some strains, electroporation can provide better results. In this case, DNA from the restriction–ligation mix should first be ethanol-precipitated and resuspended in 10 μL of water.
11. Two colonies should be sufficient in most cases, since the majority of colonies should contain the correct fragment. Clones with a correct restriction pattern do not necessarily need to be sequenced since constructs are normally assembled only from sequenced level 0 modules.
12. When libraries are made rather than individual constructs, an aliquot of the transformation is plated to evaluate the efficiency of cloning. In this case, there is usually no need to screen these colonies. The rest of the transformation is grown as a single culture. A miniprep made from this culture can be used in the same way as a miniprep from a single construct for the next stage of cloning. It can also be digested to check the quality of the library; however, the size of the fragments will of course depend on what is present in the library.
13. The optimal temperature for restriction with *Bpi*I is 37 °C. The final digestion is thus carried out at 37 °C.

References

1. Engler C, Kandzia R, Marillonnet S (2008) A one pot, one step, precision cloning method with high throughput capability. PLoS One 3: e3647
2. Weber E, Engler C, Gruetzner R, Werner S, Marillonnet S (2011) A modular cloning system for standardized assembly of multigene constructs. PLoS One 6:e16765
3. Binder A, Lambert J, Morbitzer R, Popp C, Ott T, Lahaye T, Parniske M (2014) A modular plasmid assembly kit for multigene expression,

gene silencing and silencing rescue in plants. PLoS One 9:e88218

4. Sarrion-Perdigones A, Falconi EE, Zandalinas SI, Juarez P, Fernandez-del-Carmen A, Granell A, Orzaez D (2011) GoldenBraid: an iterative cloning system for standardized assembly of reusable genetic modules. PLoS One 6:e21622
5. Sarrion-Perdigones A, Vazquez-Vilar M, Palaci J, Castelijns B, Forment J, Ziarsolo P, Blanca J, Granell A, Orzaez D (2013) GoldenBraid 2.0: a comprehensive DNA assembly framework for plant synthetic biology. Plant Physiol 162:1618–1631
6. Lampropoulos A, Sutikovic Z, Wenzl C, Maegele I, Lohmann JU, Forner J (2013) GreenGate—a novel, versatile, and efficient cloning system for plant transgenesis. PLoS One 8:e83043
7. Emami S, Yee MC, Dinneny JR (2013) A robust family of Golden Gate Agrobacterium vectors for plant synthetic biology. Front Plant Sci 4:339
8. Werner S, Engler C, Weber E, Gruetzner R, Marillonnet S (2012) Fast track assembly of multigene constructs using Golden Gate cloning and the MoClo system. Bioengineered bugs 3:38–43
9. Terfrüchte M, Joehnk B, Fajardo-Somera R, Braus GH, Riquelme M, Schipper K, Feldbrugge M (2014) Establishing a versatile Golden Gate cloning system for genetic engineering in fungi. Fungal Genet Biol 62:1–10
10. Kakui Y, Sunaga T, Arai K, Dodgson J, Ji L, Csikasz-Nagy A, Carazo-Salas R, Sato M (2015) Module-based construction of plasmids for chromosomal integration of the fission yeast Schizosaccharomyces pombe. Open Biol 5:150054
11. Lee ME, DeLoache WC, Cervantes B, Dueber JE (2015) A highly characterized yeast toolkit for modular, multipart assembly. ACS Synth Biol 4:975–986
12. Mitchell LA, Chuang J, Agmon N, Khunsriraksakul C, Phillips NA, Cai Y, Truong DM, Veerakumar A, Wang Y, Mayorga M et al (2015) Versatile genetic assembly system (VEGAS) to assemble pathways for expression in S. cerevisiae. Nucleic Acids Res 43:6620–6630
13. Agmon N, Mitchell LA, Cai Y, Ikushima S, Chuang J, Zheng A, Choi WJ, Martin JA, Caravelli K, Stracquadanio G et al (2015) Yeast golden gate (yGG) for the efficient assembly of S. cerevisiae transcription units. ACS Synth Biol 4:853–859
14. Iverson SV, Haddock TL, Beal J, Densmore DM (2016) CIDAR MoClo: improved MoClo assembly standard and new E. coli part library enable rapid combinatorial design for synthetic and traditional biology. ACS Synth Biol 5:99–103
15. Moore SJ, Lai HE, Kelwick RJ, Chee SM, Bell DJ, Polizzi KM, Freemont PS (2016) EcoFlex: a multifunctional MoClo kit for E. coli synthetic biology. ACS Synth Biol 5:1059–1069
16. Marillonnet S, Werner S (2015) Assembly of multigene constructs using golden gate cloning. Methods Mol Biol 1321:269–284
17. Engler C, Youles M, Gruetzner R, Ehnert TM, Werner S, Jones JD, Patron NJ, Marillonnet S (2014) A golden gate modular cloning toolbox for plants. ACS Synth Biol 3:839–843
18. Engler C, Marillonnet S (2014) In: Valla S, Lale R (eds) DNA cloning and assembly methods, vol 1116. Humana Press, Clifton, pp 119–131

Chapter 8

Combinatorial Assembly of Multigene Pathways by Combining Single-Strand Assembly with Golden Gate Assembly

D. Bauwens, P. Coussement, J. Maertens, and Marjan De Mey

Abstract

Biotechnological production routes for fine and bulk chemicals are progressively explored and developed. Yet this development is hampered by the many constraints determining the metabolic sweet spot, such as optimal expression levels, metabolic stress, feedback regulation, etc.

In this regard, we introduce a novel, highly reliable, and rapid single-strand assembly (SSA) methods for combinatorial pathway engineering. In this contribution, SSA is elucidated which enables one to modulate the expression via promoter and/or RBS randomization. Moreover, a new combinatorial multigene pathway assembly scheme based on single-strand assembly (SSA) methods and Golden Gate Assembly is introduced, exploiting the strengths of both assembly techniques.

Key words Single-strand assembly, Combinatorial pathway optimization, Lycopene, *E. coli*

1 Introduction

For an increasing number of fine and bulk chemicals, economically viable biotechnological production routes are being developed [1–3]. In this context, production hosts, such as *Escherichia coli* and *Saccharomyces cerevisiae*, are being transformed into true microbial cell factories, converting cheap and renewable feedstocks into high added-value products with high yields, titers and equally high productivity [4]. These microbial cell factories are provided with, often hard-to-express, multistep product pathways and their metabolism is extensively optimized in order to adequately supply the various pathway precursors [4]. This optimization typically comprises extensive expression modulation of the pathway genes involved. In this regard, massive or imbalanced expression can result in metabolic burden, caused by the excessive withdrawal of the cell's essential building blocks, reducing the cell's viability and its productivity [5]. It can also result in improper protein folding

Christine Nicole S. Santos and Parayil Kumaran Ajikumar (eds.), *Microbial Metabolic Engineering: Methods and Protocols*, Methods in Molecular Biology, vol. 1927, https://doi.org/10.1007/978-1-4939-9142-6_8,

and aggregation [6], or even in the accumulation of (toxic) pathway intermediates possibly triggering a series of regulatory events [7], adversely affecting the flux through the pathway. Hence, to avoid the detrimental effects of pathway expression, extensive fine-tuning of pathway expression with respect to the host's metabolism is of capital importance [8, 9]. In this respect, metabolic engineers dispose of, for example, promoter and ribosome binding site libraries, with varying transcription and translation initiation rates, to control the respective processes [10–14].

In order to obtain an optimally balanced enzyme expression, combinatorial assembly of DNA sequences is increasingly used. Although several methods have been developed for combinatorial pathway engineering, none of them truly allow for fast, standardized, and precise assembly to enable systematic searches of the multivariate pathway optimization landscape [15]. Moreover, such methods for pathway optimization should be amenable to automation of the synthesis and assembly processes [15].

In this context, we have previously introduced two methods: (1) the single-strand assembly (SSA) method and (2) the direct combinatorial pathway optimization method. The SSA method comprises several methods for introducing variability in double stranded DNA (dsDNA). These methods are fast, capable of standardization and easily automatable. The generalizability and reliability of these methods makes them indispensable tools for metabolic engineering [15]. The direct combinatorial pathway optimization method combines SSA with Golden Gate Assembly (GGA), exploiting the strengths of both assembly techniques for the directed combinatorial optimization of multigene pathways [15–17].

In this contribution, we describe the detailed practical implementation to perform single-strand assembly (SSA) and direct combinatorial pathway optimization.

2 Materials

All solutions and mixes are prepared using ultrapure Milli-Q water (prepared by purifying deionized water, to attain a sensitivity of 18 MΩ-cm at 25 °C). DNA fragments and plasmids are stored at −20 °C. Strains are stored in 35% glycerol at −80 °C.

2.1 Single-Strand Assembly (SSA)

2.1.1 Cloning the Backbone and Fluorescent Proteins

1. Specific oligonucleotides are designed and ordered for amplifying the plasmid backbone that contains a fluorescent protein, with a homology overlap of around 20–40 bp, *see* Fig. 1 (*see* **Note 1**).
2. A second set of oligonucleotides are designed carrying the desired randomization and the 20–40 bp overlap with the backbone as designed in **item 1**.

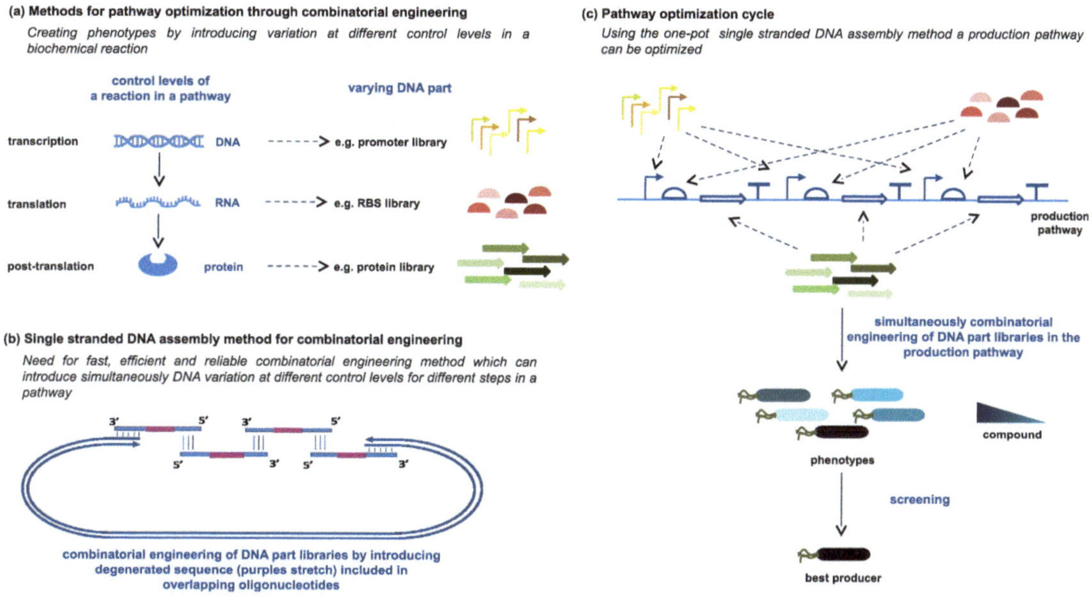

Fig. 1 Schematic overview to improve pathway optimization: (**a**) several levels of molecular control in a pathway, (**b**) schematic overview of single stranded DNA assembly (SSA), and (**c**) schematic overview of applying SSA method to reduce the pathway optimization cycle time. Reproduced from [15] with permission from Elsevier

3. Luria–Bertani (LB) medium is used for cultivating strains: 10 g of tryptone, 5 g yeast extract, 10 g sodium chloride per liter, supplemented with 30 μg/ml chloramphenicol, 25 μg/ml kanamycin, or 100 μg/ml ampicillin when necessary.
4. A Qiagen mini prep kit is used for plasmid recovery.
5. High fidelity PCR (PrimeSTAR-HS premix) is used with appropriate primers for backbone PCR amplification (*see* **Note 2**).
6. Generated fragments are agarose gel-purified (*see* **Note 3**) using a QiaQuick Gel Extraction kit and DNA concentrations are measured using a NanoDrop 2000.

2.1.2 Generating Libraries, Single-Strand Assembly

SSA mixes are composed depending on the used technique:

1. CPEC technique [18]: 100 ng backbone DNA, 400 mM of oligo(s), 5 μl Q5 reaction buffer, 5 μl of 2 mM dNTPs, 0.75 μl DMSO, 0.5 μl of 2 U/μl Q5 polymerase, and MilliQ water until a total volume of 25 μl are added in a PCR tube.
2. Gibson technique [19]: 100 ng backbone DNA, 400 mM of oligo(s), together with 15 μl of Gibson mix (320 μl ISO buffer (3 ml Tris–HCl pH 7.5, 150 μl 2 M $MgCl_2$, 240 μl 100 mM dNTP mix, 300 μl 1 M DTT, 1.5 g PEG 8000, 300 μl NAD, and Milli-Q water until a total volume of 6 ml), 0.64 μl of

10 U/μl T5 exonuclease, 20 μl of 2 U/μl phusion polymerase, 160 μl of 40 U/μl Taq ligase, and Milli-Q water until a total volume of 1.2 ml), and Milli-Q water until a total volume of 20 μl are added in a PCR tube.

2.1.3 Transformation and Recuperation of the Library

1. An *E. coli* K12 MG1655 strain is used for all purposes.
2. LB medium is used as growth medium. LB agar plates, LB medium supplemented with 10 g/l agar, are used for plating out.
3. Ice-cold Milli-Q water is used to make the cells competent.
4. Transformation mixes are composed of 50 μl competent cells and 2 μl of SSA mix.
5. A Gene Pulser Electroporator and Gene Pulser 0.1 cm gap sterile electroporation cuvettes are used to perform electroporation.
6. SOC medium, 20 g tryptone, 5 g yeast extract, 0.5 g NaCl, 2.5 ml of 1 M KCl adjusted to pH 7 with NaOH, and 10 ml of 1 M $MgCl_2$ after autoclavation per liter, is used as regeneration medium.
7. Plasmid extraction is performed using a Qiagen Miniprep kit.
8. Sequencing services are performed by Macrogen sequencing.

2.1.4 High-Throughput Fluorescence Measurement Techniques

1. A QPix2 colony picker is used for selecting individual colonies.
2. Greiner 96-well flat-bottomed microtiter plates containing 150 μl LB medium per well, supplemented with the appropriate antibiotic, are used for cultivation enclosed with a sterile Breathe-Easy sticker.
3. A Tecan M200 infinite PRO is used for measuring optical densities at 600 nm and quantification of the fluorescence output.

2.2 Direct Combinatorial Pathway Optimization

Since the direct combinatorial pathway optimization is an extension of the SSA method, previous requirements, except for the high-throughput measurement techniques, are identical. For the pathway optimization, however, some additional affairs are required (*see* **Note 4**). Structure and terminology of this method is explained in Fig. 2.

2.2.1 Combinatorial Pathway Assembly

1. LB medium is used as growth medium.
2. Specific oligonucleotides are designed and ordered for amplifying the lycopene genes and the associated carrier plasmid (pCP)-backbones (*see* Fig. 2), both containing a homology overlap of around 20–40 bp for assembly.

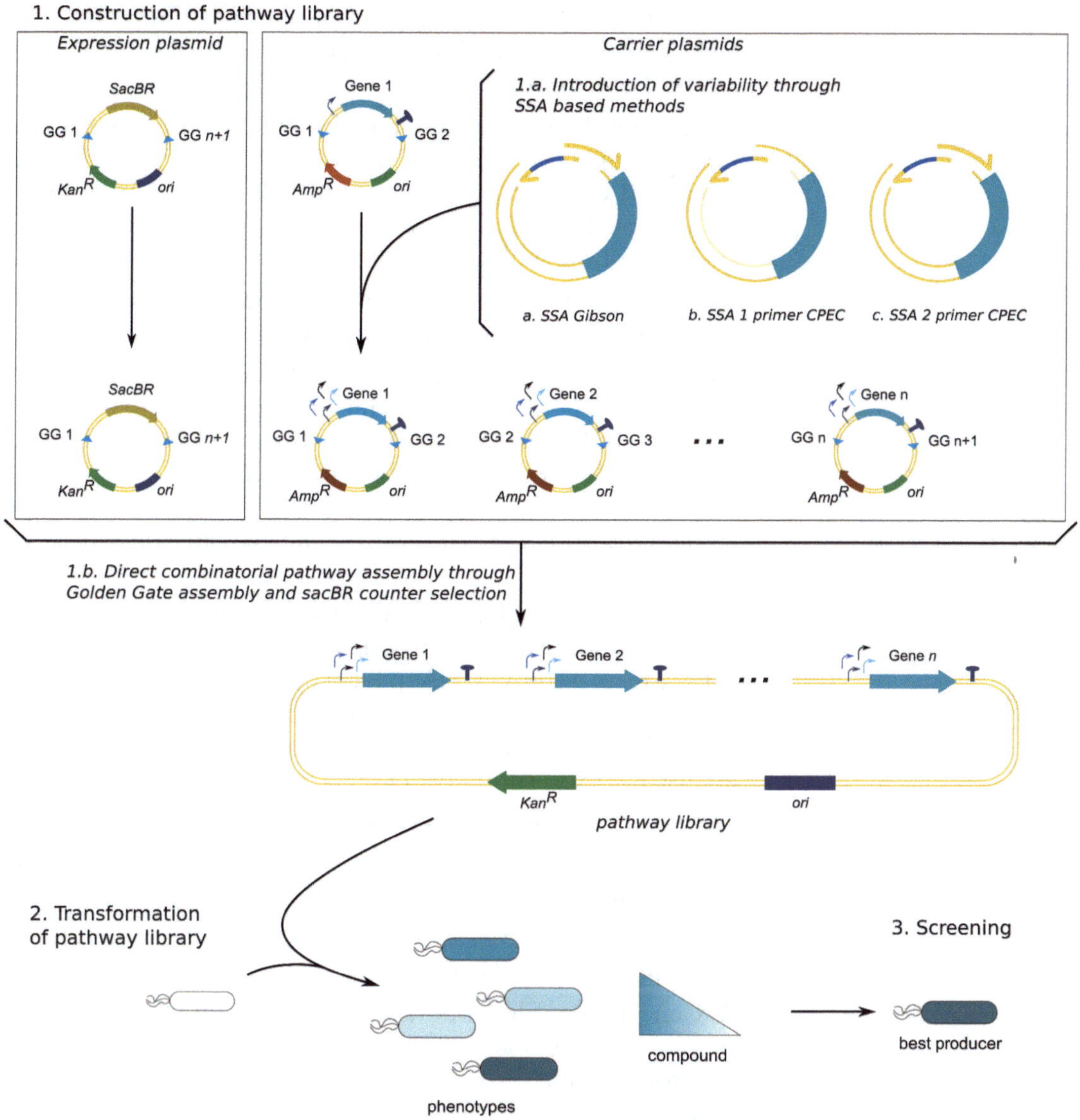

Fig. 2 Overview of the complete workflow for direct combinatorial pathway engineering. In the first step (1.a) sequence variation is introduced in each gene (n) of the pathway. In step (1.b) these individual libraries are combined into a single expression plasmid. Subsequent transformation (2) and screening (3) retrieves the strain with the desired phenotype, that is, carrying the combinatorial optimized pathway. Reproduced with permission from [16]. Copyright 2016 American Chemical Society

3. The pAC-Lyc plasmid [20] is used as a template for amplifying the individual genes for the lycopene pathway (*crtE*, *crtI*, and *crtB*).
4. High fidelity PCR (PrimeSTAR-HS premix) is used with appropriate primers for amplification of the lycopene genes and backbones.

5. Generated fragments are agarose gel-purified using a QiaQuick Gel Extraction kit and DNA concentrations are measured using a NanoDrop 2000.
6. CPEC mix: 100 ng backbone DNA, 400 mM of oligo(s), 5 μl Q5 reaction buffer, 5 μl of 2 mM dNTPs, 0.75 μl DMSO, 0.5 μl of 2 U/μl Q5 polymerase, and Milli-Q water until a total volume of 25 μl are added in a PCR tube and is used for assembly of the individual pCP plasmids.
7. SSA-mixes are prepared for promoter randomization of the three carrier plasmids, as described in Subheading 2.1.
8. Golden Gate mix is composed of 100 ng expression plasmid, equimolar amounts of randomized carrier plasmids, 1.5 μl NEB T4 Buffer, 0.15 μl BSA, 1 μl BsaI, 1 μl NEB T4 ligase, and Milli-Q water until a volume of 15 μl.

2.2.2 Transformation

1. *E. coli* One Shot TOP10 Electrocomp cells are solely used for cloning purposes.
2. An *E. coli* K-12 MG1655 wild-type strain, and two strains optimized for lycopene production (*E. coli* K-12 MG1655 DE3 ΔrecA ΔendA ΔaraA::pTrc-MEP (Trc-MEP) and *E. coli* K-12 MG1655 DE3 ΔrecA ΔendA ΔaraA::pT7-MEP (T7-MEP), where MEP stands for the dxs-idi-ispDF operon) are used as expression host [21].
3. LB medium is used as growth media.
4. Ice-cold Milli-Q water is used in order to make the cells competent.
5. Transformation mixes are composed of 50 μl competent cells and 2 μl of Golden Gate mix.
6. A Gene Pulser Electroporator and Gene Pulser 0.1 cm gap sterile electroporation cuvettes are used to perform electroporation.
7. SOC medium is used as regeneration medium.
8. LB devoid of NaCl and supplemented with 10 g/l agar, 25 μg/ml kanamycin, and 5% (m/v) sucrose is used for plating out the regeneration broth and picking colonies.

2.2.3 High-Throughput Lycopene Bioassay

1. A QPix2 colony picker is used for selecting individual colonies.
2. Polypropylene flat-bottom microtiter plates containing 150 μl LB medium per well, supplemented with the appropriate antibiotic and 0.1 mM IPTG, are used for cultivation enclosed with a sterile Breathe-Easy sticker.
3. A Tecan M200 infinite PRO is used measuring optical densities at 600 nm for biomass correction and at 443, 471, and 502 nm for lycopene quantification.

4. A Rotica swing bucket centrifuge is used for pelleting the MTPs.
5. Extraction solution is composed of 50% tetrahydrofuran (THF), 20% acetone, 30% glycerol, and 0.1% alpha-tocopherol.
6. A Lycopene standard curve is made in the extraction solution with pure commercial lycopene.

3 Methods

3.1 Single-Strand Assembly (SSA)

In function of the study's objective, SSA can be conducted with one, two, or more oligonucleotides (*see* Fig. 1). In this example, two oligonucleotides are used for randomization of the promoter and the 5′-UTR region controlling the expression of a fluorescent protein.

3.1.1 Cloning the Libraries with Fluorescent Proteins

1. Computer and software (CloneManager, Vector editor, or other) are used for in silico design and in silico cloning.
2. Oligos for picking up the backbone (*see* **Note 1**) and oligo (s) that carry the randomization and the designed overlap with the backbone are designed and ordered.
3. Strains containing the plasmid that are subject for randomization are cultivated overnight in 5 ml selective liquid media at 37 °C while shaking, before the plasmid is extracted using the proper Qiagen kit.
4. Backbone(s) are amplified from the plasmid(s) using the primeSTAR-HS premix protocol (*see* **Note 2**), shortly; the temperature profile started with 30 s at 98 °C. Next are 30 cycles of 10 s at 98 °C, 5 s at 55 °C, and 60 s per kb at 72 °C. The final step is 5 min at 72 °C before storing at 4 °C.
5. The resulting backbones are checked in length and purity by agarose gel electrophoresis and are subsequently subjected to gel-extraction purification using the proper Qiagen kit, to get rid of any residual template plasmid or false priming products (*see* **Note 3**).

3.1.2 Generating Libraries, Single-Strand Assembly

Libraries are created using different techniques, following different protocols. In a first step, the mixes are prepared. Large quantities of these mixes can be made and stored at −20 °C (*see* **Note 5**). When mixes are used from the freezer, the mixes should be thawed on ice before use. Both protocols start with the heating of the two or more oligos to 98 °C for 5 min before adding them to the respective mixes (*see* **Note 6**) and starting the temperature profile:

1. The CPEC temperature profile started with 30 s at 98 °C. Next are 15 cycles of 10 s at 98 °C, 30 s at 55 °C, and 30 s per kb at 72 °C. The final step is 5 min at 72 °C before storing at 4 °C.
2. The Gibson technique follows an isothermal protocol of 1 h at 50 °C before storing at 4 °C.

3.1.3 Transformation and Recuperation of the Library

1. An *E. coli* K-12 MG1655 strain is used for cloning and expression purposes.
2. 5 ml of LB media is inoculated with the *E. coli* strain and is grown overnight at 37 °C while shaking.
3. The overnight strain is made competent by centrifugation and disposal of the supernatant and washing a few times with ice-cold MQ water. Ideally, these steps are to be performed swiftly and on ice in order to prevent cell lysis.
4. Transformation mixes are made in prechilled electroporation cuvettes on ice.
5. After the electric pulse (1.75 kV at 200 Ω), 950 μl of SOC medium is added and the cells are transferred to a 15 ml falcon tube in which the cells are regenerated for 1 h at 37 °C while shaking.
6. After regeneration, the entire 1 ml suspense is used to inoculate 15 ml fresh LB medium supplemented with the desired antibiotics for selection and is cultivated for 24 h at 30 °C while shaking.
7. After 24 h cultivation, the library is extracted using the proper Qiagen kit, and the concentration is determined using a NanoDrop (*see* **Note 7**).
8. Finally the library is verified via sequencing (using an appropriate primer, following and ensuring sequencing requirements).

3.1.4 Transformation of the Library in the Expression Host

1. Once the library is verified, it can be transformed in the proper host (here *E. coli* K-12 MG1655). The transformation protocol for this is identical except for the fact that after regeneration, a fraction of the suspense (diluted if needed) is plated out on selective plates and incubated at 30 °C for 24 h, from where individual colonies can be analyzed and the library can be characterized (*see* **Note 8**).

3.1.5 Library Evaluation and High-Throughput Fluorescence Measurement Techniques

1. The success of the SSA method is evaluated based on sequencing results, on the numbers of colony forming units (CFUs) and the generated variation in fluorescence, which corresponds to the size of the library and the introduction of genetic variability, respectively.
2. CFUs are counted on selective plates after 24 h of cultivation at 30 °C.

3. Variation in fluorescence is determined by using a colony-picker (*see* **Note 9**) for random picking colonies, growing them in microtiter plates, and measuring the fluorescent output, quantifying the amount of fluorescent protein and correcting it for the amount of biomass (OD_{600}).

3.2 Direct Combinatorial Pathway Optimization

For the direct combinatorial pathway optimization, the SSA step needs to be performed in carrier plasmids (pCPs) containing the individual genes, flanked by the unique Golden Gate sites that represent the consecution of assembly. Finally, the pathway will be assembled in a final expression plasmid (pEX), that harbors the Golden Gate sites (*see* **Note 10**) that effectuate complementary sticky ends to the first and last gene of the pathway, flanking a counter selection marker (*see* Fig. 2). Concerning the lycopene pathway, the gene libraries for *crtE*, *crtI*, and *crtB* are created and finally assembled two different pEXs.

3.2.1 Combinatorial Pathway Assembly

1. Before starting, the whole assembly is checked in silico; the individual SSA steps within the carrier plasmids and the final Golden Gate assembly in the expression plasmid (*see* **Note 11**).
2. In a first step, the individual genes of the lycopene pathway are amplified from the carotenoid pathway gene cluster of *Erwinia herbicola* (*crtEIB*) from pAC-Lyc using high-fidelity PCR.
3. The individual genes are cloned in individual pCPs, that harbor the correct Golden Gate sites for correct assembly, via CPEC assembly.
4. Subsequently, SSA is performed as described above in the individual pCPs, in the case of lycopene production, three promoter libraries are created in front of the three genes of the pathway (i.e., *crtE*, *crtI*, and *crtB*) by randomizing the 17-nt spacer sequence between the −10 and the −35 box.
5. A final expression plasmid (pEX) is accessible (*see* **Note 12**), containing the fitting sticky ends to assemble the complete pathway in between, flanking a counter selection marker (*sacRB*). In the case of the lycopene pathway, two pEXs are used, both containing a different origin of replication, that is, the low-copy pSC101 (~5 copies) and the medium-copy pMB1 (~15–20 copies) (*see* **Note 13**).
6. Once the libraries are constructed and verified via sequencing, the individual pCP libraries are assembled into a single pEX via Golden Gate assembly, in short, the Golden Gate mix is prepared and subject to the following temperature profile; 50 cycles of alternating 5 min at 37 °C and 5 min at 16 °C, followed by 5 min at 50 °C and finally 5 min at 80 °C before storing at 4 °C.

3.2.2 Transformation of the Final Pathway

1. Transformation of the individual pCP libraries are performed in *E. coli* One Shot TOP10 Electrocomp cells, which are commercially available.
2. As an expression host, the *E. coli* K-12 MG1655 wild-type strain, the Trc-MEP and the T7-MEP strain are inoculated in 5 ml of LB and grown overnight at 37 °C while shaking.
3. The overnight strain is made competent by centrifugation and disposal of the supernatant and washing a few times with ice-cold MQ water. These steps are to be performed swiftly and on ice in order to prevent cell lysis.
4. Transformation mixes are made in prechilled electroporation cuvettes on ice.
5. After the electric pulse (1.75 kV at 200 Ω), 950 μl of SOC medium is added and the cells are transferred to a 15 ml falcon tube in which the cells are regenerated for 1 h at 37 °C while shaking.
6. After regeneration, a fraction of the suspense (diluted if needed) is plated out on selective plates (containing the counter selection) and incubated at 30 °C for 24 h.

3.2.3 Library Evaluation and High-Throughput Lycopene Bioassay

1. The success of the SSA methods and the Golden Gate assembly is evaluated on the numbers of colony forming units (CFUs) and the amount of produced lycopene, which corresponds to the size of the library and the introduction of genetic variability, respectively.
2. CFUs are counted on selective plates after 24 h of cultivation at 30 °C.
3. For both libraries in the different expression hosts, 288 colonies per library are randomly picked by an automated colony-picker and cultivated in polypropylene microtiter plates (*see* **Note 14**) in the presence of the proper inducer (IPTG) for the inducible strains, and incubated at 30 °C for 48 h while shaking.
4. After incubation, OD_{600} is measured for biomass correction.
5. The plates are cooled (4 °C) and centrifuged at least 3500 × *g* for 10 min.
6. The supernatant is discarded by inverting the plate and by gently tapping the bottom.
7. The pellet is resuspended in 150 μl extraction solution which also acts as a lysing solution (*see* **Note 15**).
8. After incubation for 10 min at 50 °C, the solution is cooled and centrifuged as described above, and 100 μl of the red colored supernatant is transferred to a fresh polypropylene 96-well plate for direct absorbance measurements.

9. Lycopene production is quantified based on its three characteristic absorbance wavelengths (443, 471, and 502 nm) using a standard curve.

4 Notes

1. Backbone means the entire plasmid except for the part that needs to be randomized (promoter, RBS, CDS, *see* Fig. 1).
2. Dilute plasmids at least 1000× to circumvent any background in fluorescent measurements.
3. In order to completely annul any background, a DpnI digest on the PCR samples before gel purification is highly recommended for at least 1 h at 37 °C.
4. Make sure that every step in the direct combinatorial pathway optimization workflow is checked in silico before performing initial SSA steps.
5. Stocks can be made if stored properly at −20 °C. However, a small drop in efficiency is observed when using frozen mixes, compared to freshly made ones.
6. If only one oligo is used, the melting step at 98 °C can be skipped.
7. Concentration and purity of the extracted library is determined on a NanoDrop, since both are important requirements for sequencing services and important parameters for determining library construction efficiency.
8. If the library is verified via sequencing, the original SSA-mix can be retransformed into the proper expression host (if stored correctly, at −20 °C). Otherwise, the extracted library can be used for transformation.
9. Colonies could be picked by hand. However, we strongly recommend using an automated colony picking device for picking large amounts of colonies to ensure completely random picking and to reduce the amount of human errors.
10. When designing the Golden Gate step for a lengthy pathway, some adjustments have to be taken into account for optimizing the efficiency of the assembly [17], (1) using a high-fidelity restriction enzyme (e.g., BsaI-HF), (2) avoiding multiple complementary and palindromic sequences within sticky-ends, (3) sticky-ends should differ by at least two nucleotides, (4) sticky-ends should have a GC-content of 50%, and finally (5) using vectors (pCPs and pEXs) that are Golden Gate proof (the backbone should not contain any extra restriction sites).
11. When designing the complete direct combinatorial pathway optimization experiment, consider designing a general work-

flow rather than a single experiment, since a standardized set of plasmids (*see* Fig. 2) will allow for maximal reusage of primers (degenerated and nondegenerated), carrier and expression plasmids (pCPs and pEXs).

12. In order to enhance selection of successful constructs, pEX(s) should harbor a different antibiotic resistance marker than pCPs.
13. Illustrating the ease of use and the commutativity of the standardized set of plasmids, the lycopene pathway is combinatorially optimized in parallel, in two different expression vectors, using the same libraries.
14. Microtiter plates manufactured from polypropylene are used here because of their inherent resistance to tetrahydrofuran (THF), which is the main component of the extraction solution.
15. Since the extraction mix contains solvents (THF and acetone), appropriate precautions should be taken. Wear appropriate gloves (nitrile or neoprene) and work in an appropriate environment (fume hood).

References

1. Jullesson D, David F, Pfleger B, Nielsen J (2015) Impact of synthetic biology and metabolic engineering on industrial production of fine chemicals. Biotechnol Adv 33:1395
2. Fisher AK, Freedman BG, Bevan DR, Senger RS (2014) A review of metabolic and enzymatic engineering strategies for designing and optimizing performance of microbial cell factories. Comput Struct Biotechnol J 11:91–99
3. Choi S, Song CW, Shin JH, Lee SY (2015) Biorefineries for the production of top building block chemicals and their derivatives. Metab Eng 28:223–239
4. Keasling JD (2010) Manufacturing molecules through metabolic engineering. Science 330:1355–13–8
5. Jones KL, Kim S-W, Keasling J (2000) Low-copy plasmids can perform as well as or better than high-copy plasmids for metabolic engineering of bacteria. Metab Eng 2:328–338
6. Kopito RR (2000) Aggresomes, inclusion bodies and protein aggregation. Trends Cell Biol 10:524–530
7. Seo SW, Yang J-S, Cho H-S, Yang J, Kim SC, Park JM, Kim S, Jung GY (2014) Predictive combinatorial design ofmRNA translation initiation regions for systematic optimization of gene expression levels. Sci Rep 4:4515. https://doi.org/10.1038/srep04515
8. Neubauer P, Lin HY, Mathiszik B (2003) Metabolic load of recombinant protein production: inhibition of cellular capacities for glucose uptake and respiration after induction of a heterologous gene in Escherichia coli. Biotechnol Bioeng 83:53–64
9. Pitera DJ, Paddon CJ, Newman JD, Keasling JD (2007) Balancing a heterologous mevalonate pathway for improved isoprenoid production in Escherichia coli. Metab Eng 9:193–207
10. De Mey M, Maertens J, Lequeux GJ, Soetaert WK, Vandamme EJ (2007) Construction and model-based analysis of a promoter library for E. coli: an indispensable tool for metabolic engineering. BMC Biotechnol 7:34
11. De Mey M, Maertens J, Boogmans S, Soetaert WK, Vandamme EJ, Cunin R, Foulquié -Moreno MR (2010) Promoter knock-in: a novel rational method for the fine tuning of genes. BMC Biotechnol 10:26
12. Mutalik VK, Guimaraes JC, Cambray G, Lam C, Christoffersen MJ, Mai Q-A, Tran AB, Paull M, Keasling JD, Arkin AP, Endy D (2013) Precise and reliable gene expression via standard transcription and translation initiation element. Nat Methods 10:1–42
13. Salis HM, Mirsky EA, Voigt CA (2009) Automated design of synthetic ribosome binding

sites to control protein expression. Nat Biotechnol 27:946–950
14. Leonard E, Ajikumar PK, Thayer K, Xiao W-H, Mo JD, Tidor B, Stephanopoulos G, Prather KLJ (2010) Combining metabolic and protein engineering of a terpenoid biosynthetic pathway for overproduction and selectivity control. Proc Natl Acad Sci U S A 107:13654–13659
15. Coussement P, Maertens J, Beauprez J, Van Bellegem W, De Mey M (2014) One step DNA assembly for combinatorial metabolic engineering. Metab Eng 23:70–77
16. Coussement P, Bauwens D, Maertens J, De Mey M (2016) Direct combinatorial pathway optimization. ACS Synth Biol 6(2):224–232. https://doi.org/10.1021/acssynbio.6b00122
17. Engler C, Marillonnet S (2011) Generation of families of construct variants using golden gate shuffling. Methods Mol Biol 729:167–181
18. Quan J, Tian J (2009) Circular polymerase extension cloning of complex gene libraries and pathways. PLoS One 4:e6441
19. Gibson DG, Young L, Chuang R-y Y, Venter JC, Iii CAH, Smith HO, America N (2009) Enzymatic assembly of DNA molecules up to several hundred kilobases. Nat Methods 6:12–16
20. Cunningham FX, Sun Z, Chamovitz D, Hirschberg J, Gantt E (1994) Molecular structure and enzymatic function of lycopene cyclase from the cyanobacterium Synechococcus sp strain PCC7942. Plant Cell 6:1107–1121
21. Ajikumar PK, Xiao W-H, Tyo KEJ, Wang Y, Simeon F, Leonard E, Mucha O, Phon TH, Pfeifer B, Stephanopoulos G (2010) Isoprenoid pathway optimization for Taxol precursor overproduction in *Escherichia coli*. Science 330:70–74

Chapter 9

Checks and Balances with Use of the Keio Collection for Phenotype Testing

Sandra J. Aedo, Helena R. Ma, and Mark P. Brynildsen

Abstract

The Keio single gene knockout collection comprises approximately 4000 mutants of *E. coli* K12 strain BW25113, where each mutant contains a kanamycin resistance cassette in place of a single nonessential gene. This mutant library has proven to be incredibly useful in the fields of bacteriology, chemical genomics, biotechnology, and systems biology, which is evidenced by the greater than 3800 citations that the article describing its construction has garnered in the approximate first 11 years since its publication. Among the various applications of the collection, the most extensive use has been in the assessment of how loss of specific gene function influences phenotypes. In this chapter, we describe pitfalls with use of the collection and procedures that can be employed to ensure robust phenotype assessment of mutations in the library. These include procedures for thorough confirmation of gene deletions by PCR, phage transduction of mutated loci to new host strains, and strategies for genetic complementation.

Key words *Escherichia coli*, Knockout, P1 phage transduction, Gene duplication

1 Introduction

Introduced in 2006, the Keio collection consisted of 3985 single-gene deletion mutants for nonessential genes in *E. coli* K-12 strain BW25113 [1]. Generally, these mutants were made by replacing the target gene sequence, spanning from the second to the seventh to the last codon based on the annotated ORFs at the time of the library construction, with a kanamycin resistance (kan^R) cassette using the λ-red method of Datsenko and Wanner [1, 2]. Notably, the kan^R cassette contained flanking FLP recognition target (FRT) sites that were incorporated to allow for excision of the majority of the cassette with expression of FLP recombinase. Upon excision of the kan^R cassette, an 81-nt sequence, referred to as scar sequence, is

Sandra J. Aedo and Helena R. Ma contributed equally to this work.

Christine Nicole S. Santos and Parayil Kumaran Ajikumar (eds.), *Microbial Metabolic Engineering: Methods and Protocols*, Methods in Molecular Biology, vol. 1927, https://doi.org/10.1007/978-1-4939-9142-6_9,

generated (*see* **Note 1**). This scar sequence is in-frame with the start codon and the last seven codons of the target gene so as to prevent polar effects when the target gene is part of an operon (*see* **Note 2**). The collection is available as individual mutant strains for single-gene analysis from the National Institute of Genetics of Japan, The Coli Genetic Stock Center at Yale, and Open Biosystems. A 96-well microtiter plate (45 plates) format for high throughput purposes is available as well from the National Institute of Genetics of Japan and Open Biosystems.

The Keio collection is a valuable widespread resource, and according to an update on the Keio collection in 2009 [3] more than four million samples had been distributed worldwide since its construction in 2006. A few examples of the utility of this library include its use in the characterization of novel genes with functions in thiamine salvage [4], the identification of enzymatic functions mediating lipid modification of *E. coli* outer membrane in biofilms [5], and the identification of metabolic processes associated with persister formation [6–8]. In addition, the Keio collection has been used as a powerful high-throughput screening strategy, for example, in the identification of genes involved in antibiotic sensitivity and biofilm formation [9–11]. Further, the 96-well plate format of the Keio collection in combination with automated phage transduction has provided a way to rapidly generate collections of host strains with directed mutations [12]. These examples illustrate the vast utility that the Keio collection provides for assessing the phenotypic landscape of loss of function mutants of *E. coli*.

As with any resource, use of the Keio collection must be conducted with an appreciation of its limitations, of which several were recognized and discussed in the original article [1]. One issue was that construction of some mutants can affect the coding sequence of a neighboring gene when the ORF of the target gene overlaps with the ORF of a neighboring gene (at the time the library was constructed, 742 overlapping genes were found in *E. coli* K-12). Another issue was that 12 genes were deleted not in-frame because their sequences were reannotated after the construction of their corresponding mutants [13], and those genes are listed in Supplementary Table 5 in the original article [1]. Other problems discussed by the authors included potential handling errors, cross-contamination, the presence of secondary mutations, and possible gene duplication. The issues concerning gene duplication and cross-contamination were addressed in the update to the Keio collection in 2009 where it was conveyed that 33 mutants were cross-contaminated and 25 contained a duplication of the gene that was intended for deletion [3]. With regard to the original criteria for mutant verification, four or eight colonies were chosen from each deletion experiment for confirmation by PCR using locus- and kan^{R}-specific primers. From those PCR tests, a mutant was scored

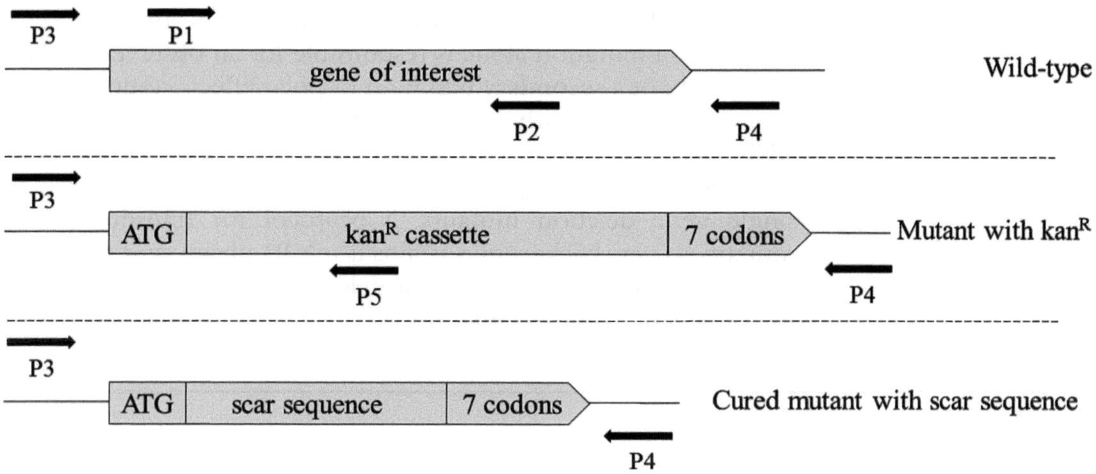

Fig. 1 Map of genomic elements and primers. A forward internal primer (P1) and a reverse internal primer (P2) to the gene of interest are used to confirm that the target gene has not been duplicated in the mutant strain. A forward external primer for the gene of interest (P3) and a reverse external primer to the gene of interest (P4) may be used to verify the expected size of the gene of interest in wild type, the kan^R cassette in the mutant, or the scar sequence in a cured mutant. A forward external primer for the gene of interest (P3) and a reverse internal primer for the kan^R cassette (P5) are used to confirm that kan^R is present in the intended chromosomal location in the mutant strain

as correct if two or more colonies had the expected structure. For a large number of mutants all colonies selected provided positive PCR results; however, there were many mutants that did not.

To assess if any of the issues mentioned above are present in a Keio mutant, several steps can be performed. Two sets of PCR reactions with the purpose of verifying that kan^R is present in the intended chromosomal location and that the target gene has not been duplicated are advisable (Fig. 1). To reduce the likelihood that secondary mutations in Keio mutants might impact phenotype assessment, the collection is commonly used in combination with phage transduction, which transmits the region of the chromosome that contains the desired mutation to a new host strain. With generalized phage transduction, the donor bacterium is infected with a transducing phage that can package some of its host bacterial DNA instead of its own DNA into its head [14]. Phage is then used to infect the recipient bacterium where transducing particles within the phage population inject donor bacterial DNA instead of phage DNA [14]. One good transducing phage is P1 phage, which can package and deliver DNA of approximately 100 kb in length [14, 15]. Because transduction of a desired chromosomal region is a rare event, selection markers are needed to identify recombinants carrying the desired chromosomal DNA [14]. With the Keio collection, transductants are selected for kanamycin resistance and a number of them are tested for the mutation by PCR. In addition, to verify that a mutation alone is responsible for an observed phenotypic, and not a secondary mutation or polar effect, complementation experiments with a strain cured of the kan^R cassette are advisable.

Here, we describe procedures for genetic confirmation of Keio single gene deletion mutants, a protocol for transduction of mutants to new background strains using P1 phage, and a strategy to confirm the importance of gene deletions to phenotypes with the use of genetic complementation.

2 Materials

Prepare all solutions using ultrapure water (deionized water purified to attain a resistivity of 18.2 MΩ·cm at 25 °C). Store at room temperature, unless otherwise specified. A 0.22 μm filter should be used for all sterile filtering.

2.1 Media

1. Luria–Bertani (LB) medium. Dissolve 10 g tryptone, 5 g yeast extract, and 10 g NaCl in 1 L H_2O. Autoclave at 121 °C for 30 min.
2. LB-agar plates, without antibiotic, with 50 μg/mL kanamycin, or with 100 μg/mL ampicillin. Add 1 L H_2O to 10 g tryptone, 5 g yeast extract, 10 g NaCl, and 15 g agar powder. Autoclave at 121 °C for 30 min. Let media cool to 50–60 °C. If making kanamycin plates, add 1 mL 50 mg/mL kanamycin stock solution (*see* Subheading 2.2, **item 1**). If making ampicillin plates, add 1 mL 100 mg/mL ampicillin stock solution (*see* Subheading 2.2, **item 2**). Pour agar into petri dishes and let solidify.
3. SOC medium. To 20 g tryptone and 5 g yeast extract add 955.5 mL H_2O and 2 mL of 5 M NaCl. Autoclave at 121 °C for 30 min. Let media cool to 50–60 °C. Add 2.5 mL of 1 M KCl, 10 mL of 1 M $MgSO_4$, 10 mL of 1 M $MgCl_2$, and 20 mL of 1 M glucose (all solutions sterile filtered).

2.2 Antibiotics

1. Kanamycin stock solution: 50 mg/mL in H_2O. Sterile filter.
2. Ampicillin stock solution: 100 mg/mL in H_2O. Sterile filter.

2.3 PCR for Knockout Verification

1. 10× Standard Taq Reaction Buffer. Store at −20 °C.
2. 2 mM equimolar dNTP Solution mix. Dilute 200 μL of 10 mM dNTP solution with 800 μL H_2O. Store at −20 °C.

3. Primer sets at 10 μM each (*see* **Note 3**) (Fig. 1):
   (a) Forward internal for gene of interest and reverse internal for gene of interest.
   (b) Forward external for gene of interest and reverse internal for kan^R cassette (*see* **Note 4**).
4. 25 mM $MgCl_2$ stock solution. Dissolve 2.38 mg $MgCl_2$ in 1 mL H_2O. Sterile filter. Can be stored at −20 °C with other PCR reagents.
5. 2× PCR mix for each primer set. Prepare 10 μL mix for each sample; scale up as needed. To make 10 μL of 2× mix for a primer set, add 5 μL H_2O to a PCR tube. Add 2 μL of 10× standard Taq reaction buffer. Add 1 μL each of 10 μM stock of forward and reverse primers. Finally, add 0.4 μL of 2 mM dNTPs and 0.6 μM of 25 mM $MgCl_2$. Make immediately before use.
6. Taq Polymerase (5000 units/mL).
7. 1× TAE Buffer.
8. Agarose.
9. 10 mg/mL ethidium bromide in H_2O.
10. DNA ladder with appropriate length resolution for expected PCR products, which are typically between 0.5 and 1.5 kb.

2.4 P1 Bacteriophage-Mediated Transduction

1. 1 M $CaCl_2$ stock solution. Dissolve 110.98 mg of $CaCl_2$ per 1 mL H_2O. Sterile filter.
2. 1 M $MgSO_4$ stock solution. Dissolve 120.37 mg of $MgSO_4$ per 1 mL H_2O. Sterile filter.
3. 1 M sodium citrate stock solution. Dissolve 258.06 mg of trisodium citrate per 1 mL H_2O. Sterile filter.
4. LB supplemented with 7 mM $CaCl_2$ + 12 mM $MgSO_4$ (7 mL per transduction). Add 1.75 mL of 1 M $CaCl_2$ and 3 mL of 1 M $MgSO_4$ to 250 mL LB medium.
5. LB + 25 mM sodium citrate + kanamycin agar plates. Make as in Subheading 2.1, **item 2**, but add 25 mL of 1 M sodium citrate stock solution at the same time as kanamycin before pouring plates.
6. P1 phage stock, generated by cultivation on the recipient wild-type strain (e.g., MG1655). Store at 4 °C.
7. Donor strain: strain containing the desired mutation to be moved into the recipient strain (e.g., Keio mutant).
8. Recipient strain: strain that desired mutation is to be moved into (e.g., MG1655) (*see* **Note 5**).
9. Chloroform.

2.5 Removal of the Antibiotic Resistance Marker

1. pCP20 plasmid [12].
2. 10% glycerol. Add 10 mL glycerol per 90 mL H_2O. Autoclave at 121 °C for 30 min.
3. Primer sets at 10 μM each:
   (a) Forward external for gene of interest and reverse external for gene of interest.
   (b) Forward internal for gene of interest and reverse internal for gene of interest.
   (c) Forward external for gene of interest and reverse internal for kanamycin.

2.6 Complementation Assay

1. Low-copy plasmid (e.g., pUA66).

2.7 Bacterial Strains

1. Keio collection [2]. The Keio collection is a set of 3985 single-deletion mutants in the parent strain *Escherichia coli* K-12 BW25113. BW25113 has the following mutations:
   (a) Δ*(araD-araB)567*.
   (b) Δ*lacZ4787*(::rrnB-3).
   (c) *λ*-.
   (d) *rph-1*.
   (e) Δ*(rhaD-rhaB)568*.
   (f) *hsdR514*.
2. MG1655 [16]. This is the strain we typically use as a recipient strain in these experiments, although other P1-compatible strains may be used instead.

3 Methods

3.1 PCR Confirmation of Knockout Mutation

1. Streak out Keio mutant onto a LB-agar + kanamycin plate and streak wild-type strain from stock onto a plate without antibiotic. Incubate at 37 °C overnight, typically for 16 h.
2. For each culture, pick one colony with a pipette tip or inoculation loop and inoculate into 1 mL LB in a test tube. For kan^R cassette-bearing mutants, add 1 μL kanamycin stock solution to achieve a final concentration of 50 μg/mL. Incubate for approximately 3 h at 37 °C with shaking (*see* **Note 6**).
3. Each strain should be subjected to PCR with both primer sets (*see* **Note 7**). For each of these samples, prepare a PCR

tube by adding 2 μL of culture to 8 μL H_2O and mixing thoroughly.

4. Enter thermocycling protocol as follows:

Step	Temperature	Time (min)
1	94 °C	5:00
2	94 °C	0:30
3	T_{anneal} (*see* **Note 8**)	0:30
4	72 °C	$t_{elongate}$ (1 min/kb)
5	Go to **Step 2**	Repeat 29 times
6	72 °C	5:00
7	Hold at 4 °C	

5. Place PCR tubes in PCR machine and begin Step 1 to lyse cells.
6. On ice, add 0.2 μL of 5000 units/mL Taq polymerase to PCR mixes (*see* Subheading 2.3, **item 5**).
7. Pause thermocycling protocol after the initial 5 min (cell lysis step). Add 10 μL of appropriate 2× PCR mixes.
8. Continue with thermocycling protocol until completion.
9. While thermocycling, make 1% agarose gel with 0.5 μg/mL ethidium bromide in TAE buffer.
   (a) In a flask, add 50 mL 1× TAE buffer to 0.5 g agarose.
   (b) Microwave for 1 min.
   (c) Remove from microwave using a thermal glove, swirl gently, and inspect for liquid homogeneity.
   (d) If heterogeneity is observed (visible specs of agar undissolved in liquid), microwave for an additional 15 s, swirl gently, and inspect again. Repeat this step until homogeneity is observed.
   (e) Let the mixture cool until the flask can be touched without a thermal glove, then add 2.5 μL of 10 mg/mL ethidium bromide.
   (f) Swirl to mix, then pour into mold and let cool until solid.
10. Add 4 μL of 6× loading dye to samples. Load gel with samples and DNA ladder and run in electrophoresis system in TAE buffer at 120 V (*see* **Note 9**).
11. Image agarose gel to visualize fluorescent ethidium bromide stains. A single band indicates the presence of the amplified sequence (Fig. 2).

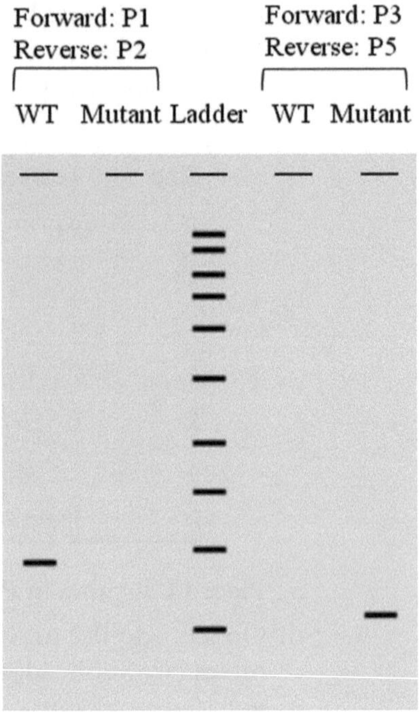

Fig. 2 Diagram of expected banding pattern for PCR on wild type and kan[R] mutant. Primers P1 and P2 yield an expected band that confirms the presence of the target gene in the wild type, and the corresponding band is absent in the knockout mutant. P3 and P5 primers set yields an appropriately sized band that confirms the proper location of the kan[R] cassette in the mutant strain, and the corresponding band is absent in the wild-type strain

3.2 P1 Bacteriophage-Mediated Transduction

3.2.1 Phage Generation (See Note 10)

1. Set up an overnight culture of the donor strain (containing the desired mutation) and the recipient strain in 1 mL LB. For a Keio mutant donor strain, add 1 μL kanamycin stock solution to achieve a final concentration of 50 μg/mL. Incubate at 37 °C for 16 h.
2. In a 15 mL conical centrifuge tube (*see* **Note 11**), add 50 μL of P1 phage stock to 50 μL of the donor strain overnight culture in 5 mL of the LB supplemented with 7 mM $CaCl_2$ + 12 mM $MgSO_4$.
3. Incubate without shaking at 37 °C for 20 min.
4. Incubate with shaking at 37 °C for 2 h.
5. Add 150 μL of the donor strain overnight culture. Repeat incubation steps.
6. Add 1 mL chloroform, then vortex for 5 s.
7. Centrifuge for 15 min at 1800 × *g*.
8. Transfer 1.4 mL of supernatant to a microcentrifuge tube and discard subnatant. Store generated phage stock at 4 °C or proceed directly to Subheading 3.2.2.

3.2.2 Phage Transduction

1. Prepare two sterile microcentrifuge tubes with 1 mL each of supplemented LB.
2. Add 400 μL of phage stock to tube 1. Pipette up and down until mixed well, then transfer 100 μL to tube 2 (*see* **Note 12**).
3. Add 100 μL of recipient strain overnight culture to each tube (*see* **Note 13**). Incubate with shaking at 37 °C for 1 h.
4. Centrifuge tubes for 90 s at 8000 × *g*.
5. Remove supernatant until 500 μL remain in each tube (*see* **Note 14**). Resuspend pellets and combine the tubes.
6. Plate 500 μL on LB-agar + sodium citrate + kanamycin plates. Spread using a sterilized cell spreader (*see* **Note 15**). Incubate at 37 °C for 16 h.
7. Restreak colonies onto LB-agar + kanamycin plates to obtain single colonies.
8. Perform PCR protocol outlined in Subheading 3.1 using the recipient strain and the newly constructed strain to verify the success of the transduction.

3.3 Removal of the kan^R Cassette

1. Set up an overnight culture of the strain in 1 mL LB. Add 1 μL kanamycin stock solution to achieve a final concentration of 50 μg/mL. Incubate at 37 °C for 16 h.
2. Centrifuge at 3000 × *g* for 5 min at 4 °C. Remove supernatant and resuspend in 750 μL 10% glycerol. Repeat twice, but on the final repetition resuspend in only 150 μL 10% glycerol.
3. Add 10–50 ng of pCP20 plasmid and incubate for 20–30 min on ice.
4. Electroporate at 2400 V. Immediately add 100 μL SOC and transfer to sterile test tube.
5. Incubate at 30 °C with shaking for 1–3 h (*see* **Note 16**).
6. Plate 500 μL onto LB-agar + ampicillin plates and incubate overnight at 30 °C.
7. Restreak colonies on LB-agar plates prewarmed to 42 °C. Incubate overnight at 42 °C.
8. Inoculate colony into 1 mL LB. Incubate at 37 °C. After culture is visibly turbid, transfer 10 μL of culture into two tubes with 1 mL LB. Add 1 μL kanamycin stock solution to the first tube to achieve a final concentration of 50 μg/mL and add 1 μL ampicillin stock solution to the second tube to achieve a final concentration of 100 μg/mL. Incubate the first tube at 37 °C to test for loss of kan^R. Incubate the second tube at 30 °C to test for the loss of the pCP20 plasmid, which confers ampicillin resistance. Incubate all three cultures overnight (*see* **Note 17**).
9. Check that the cultures did not grow in ampicillin or kanamycin (*see* **Note 18**).

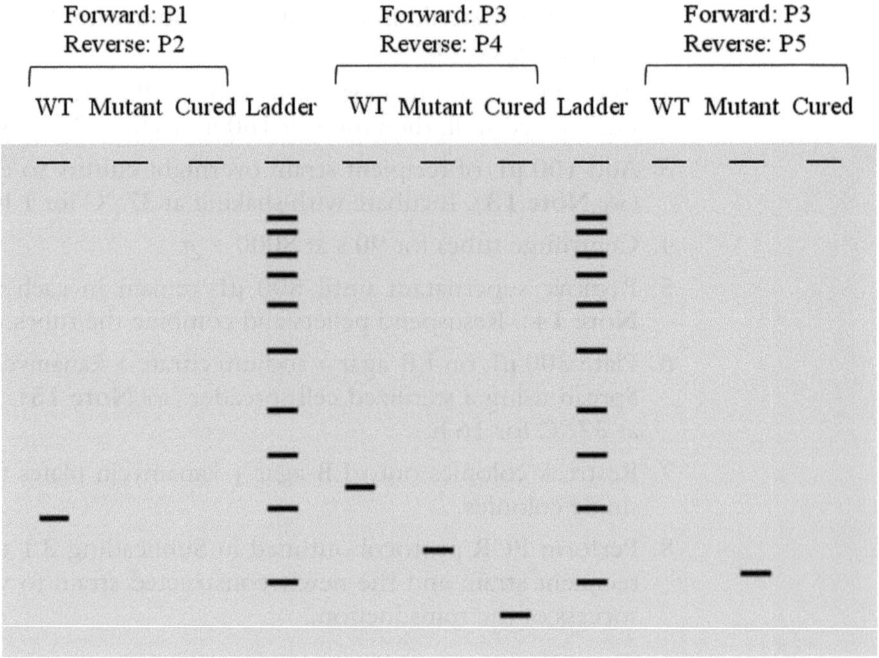

Fig. 3 Diagram of expected banding pattern of PCR products for wild type, kan^R mutant, and cured mutant. Primers P1 and P2 yield an appropriately sized band that confirms the presence of the gene of interest in the wild type, and the corresponding band is absent in the kan^R mutant and cured strain. P3 and P4 primers yields a band indicative of the presence of the gene of interest in wild type, a band indicative of the presence of the kan^R cassette in the mutant, and a band indicative of the scar sequence in the cured mutant. Primers P3 and P5 yield a band that confirms the proper location of the kan^R cassette in the mutant strain, and the corresponding band is absent in the wild-type and cured strains

10. Repeat PCR protocol outlined in Subheading 3.1 using the strain cured of kan^R, the strain with the kan^R cassette, wild type, and the three primer sets outlined in Subheading 2.5 (Fig. 3).

3.4 Complementation Strategies

Complementation assays are conducted to ensure that a given phenotype is caused by the mutation in question. If so, complementing the deletion mutant by transforming a plasmid containing the gene back into the mutant should restore the wild-type phenotype. The gene should be cloned into a plasmid and then transformed into the deletion strain; as this protocol will vary gene-to-gene it is not elaborated here. However, we recommend the following strategies for conducting complementation.

1. A major goal is to ensure that the expression level of the gene is as close to wild type as possible. To achieve this goal:
   (a) Use a low copy-number plasmid as the vector for the gene, such as pUA66 [17], which has a pSC101 origin of replication.

(b) The gene should be under the control of its native promoter.

(c) If the gene is part of an operon, the full operon can be included on the plasmid.

2. In order to assess whether complementation restored the wild-type phenotype and that the vector itself did not affect the phenotype, we recommend that the empty vector be transformed into both the wild type and the deletion mutant and the phenotypes compared to the complemented mutant. Another option is to transform the vector expressing a model protein such as GFP or LacZ from the same promoter as the gene of interest into the wild-type and deletion mutant.

4 Notes

1. The scar sequence is composed of the 28-nt sequence upstream of the FRT site 5′ to the kan^R gene, the 34-nt sequence corresponding to the FRT site sequence, and the 19-nt sequence downstream of the FRT site 3′ to the kan^R gene.
2. Excision of the kan^R cassette reduces disruption of surrounding sequences. For example, if the genetic deletion occurs in an operon and the kan^R cassette remains, it is possible that expression of the genes downstream of the cassette will be disrupted due to its transcription termination signal.
3. To confirm the absence of a gene in a knockout strain, primers internal to the gene should be used to amplify genomic DNA from the mutant and wild type. In this case, wild type should produce an appropriately sized band on an agarose gel with ethidium bromide, and the knockout mutant should not yield any band. To confirm proper location of the kan^R cassette, one primer external to the cassette insertion site and one primer internal to the cassette should be used to amplify genomic DNA from the mutant and wild type. In this case, the knockout mutant should produce an appropriately sized band on an agarose gel with ethidium bromide, whereas the wild type should not yield any band. While the exact location of primers can vary, for primers external to the gene we typically use ones located 100–200 bp outside of the start and stop codons. Primer length is typically 20–30 bp, and we seek to generate PCR products that are approximately 1–1.5 kb.
4. We typically use a reverse internal primer for the kan^R cassette and a forward external primer for the gene of interest. We use the following primer for the kan^R cassette:

   Reverse internal: 5′-GAAGCGGTCAGCCCATTC-3′.

5. Since the P1 phage packages 100 kb segments of donor DNA that will be recombined with the chromosome of the recipient, it is possible to "overwrite" existing mutations in the recipient strain if they are too close to the mutation being transduced from the donor. The situation should be assessed by checking for gene proximity between the donor and the recipient, and if proximity is a concern additional PCR checks for retention of previous mutations to the recipient strain should be conducted.
6. PCR can also be performed by skipping directly to Subheading 3.3, **step 3** by picking colonies and mixing directly with 10 μL H_2O. While this saves time, inoculating first into LB and allowing time for growth can lead to greater consistency in the amount of cells used and better mixing. In addition, preceding PCR with propagation in liquid provides a culture that can be immediately used for permanent storage or subsequent procedures, such as P1 phage transduction.
7. If the gene of interest is very different in size from the kan^R cassette, where the size difference is easily resolvable with gel electrophoresis, forward and reverse primers external to the gene of interest may also be used as a third set of primers to verify the expected size of the gene of interest in wild type and the insert in the mutant.
8. We typically use an annealing temperature of 60 °C. However, this temperature can be adjusted as necessary. For instance, it may be necessary to anneal at a lower temperature for shorter primers, or increased temperature if nonspecific bands are observed.
9. Sample volume can be as low as 3 μL per lane if presence/absence of band is the only information sought. Otherwise, sample volume can vary depending on the width and depth of lanes in the mold used. While 10 μL is often a typical loading amount, bands from high-volume samples can be gel-extracted and sequenced if additional confirmation is desired. Gels may be run for differing amounts of time, but 30–60 min is typically sufficient to confirm presence/absence of bands of expected sizes.
10. This generation step can also be used to regenerate P1 phage stock if wild-type bacteria are used as the donor strain. In our experience, phage stock will remain potent if regenerated every few months.
11. Conduct all steps involving P1 phage using sterilized disposable containers to prevent contamination of future cultures and stocks.
12. This is an effective 1:60 dilution of the phage. Conducting the experiment with multiple concentrations of phage helps ensure that at least one of the cultures will not undergo full lysis.

13. In order to verify that all donor strain bacteria were eliminated during the chloroform treatment, another check is to repeat Subheading 3.2.2 in another set of microcentrifuge tubes, adding LB in place of recipient strain. If colonies appear on plates from these tubes, donor strain is still present in the generated phage stock. Similarly, if moving genes from the Keio collection to a strain like MG1655, one can check the resulting colonies for growth on arabinose, since the BW25113 parent strain of the Keio collection cannot grow on arabinose, whereas MG1655 can.
14. A common visual check after this step is to verify that the pellet from the tube with more concentrated phage stock is smaller than the pellet from the tube with less concentrated phage stock, which suggests that phage are successfully infecting and lysing cells.
15. Other plating methods may be used here, such as plating beads, but the amount plated may need to be adjusted.
16. Some mutants may take longer to recover from electroporation than others. Adjust length of this step if necessary (for instance, if no colonies are observed).
17. Alternatively, colonies obtained following incubation at 42 °C may be replica plated on an LB-agar plate, an LB-agar plate + ampicillin, and an LB-agar plate + kanamycin.
18. If cultures grow in ampicillin but not in kanamycin, restreak to purify colony again at 42 °C and repeat test for loss of ampicillin resistance.

Acknowledgments

This work was supported by the National Institute of Allergy and Infectious Diseases of the National Institutes of Health under award number R21AI115075. The content is solely the responsibility of the authors and does not necessarily represent the official views of the funding agencies.

References

1. Baba T, Ara T, Hasegawa M, Takai Y, Okumura Y, Baba M, Datsenko KA, Tomita M, Wanner BL, Mori H (2006) Construction of *Escherichia coli* K-12 in-frame, single-gene knockout mutants: the Keio collection. Mol Syst Biol 2:2006.0008. https://doi.org/10.1038/msb4100050
2. Datsenko KA, Wanner BL (2000) One-step inactivation of chromosomal genes in *Escherichia coli* K-12 using PCR products. Proc Natl Acad Sci U S A 97(12):6640–6645. https://doi.org/10.1073/pnas.120163297
3. Yamamoto N, Nakahigashi K, Nakamichi T, Yoshino M, Takai Y, Touda Y, Furubayashi A, Kinjyo S, Dose H, Hasegawa M, Datsenko KA, Nakayashiki T, Tomita M, Wanner BL, Mori H (2009) Update on the Keio collection of *Escherichia coli* single-gene deletion mutants. Mol Syst Biol 5:335. https://doi.org/10.1038/msb.2009.92

4. Melnick J, Lis E, Park JH, Kinsland C, Mori H, Baba T, Perkins J, Schyns G, Vassieva O, Osterman A, Begley TP (2004) Identification of the two missing bacterial genes involved in thiamine salvage: thiamine pyrophosphokinase and thiamine kinase. J Bacteriol 186 (11):3660–3662. https://doi.org/10.1128/JB.186.11.3660-3662.2004
5. Chalabaev S, Chauhan A, Novikov A, Iyer P, Szczesny M, Beloin C, Caroff M, Ghigo JM (2014) Biofilms formed by gram-negative bacteria undergo increased lipid a palmitoylation, enhancing in vivo survival. MBio 5(4). https://doi.org/10.1128/mBio.01116-14
6. Orman MA, Brynildsen MP (2016) Persister formation in *Escherichia coli* can be inhibited by treatment with nitric oxide. Free Radic Biol Med 93:145–154. https://doi.org/10.1016/j.freeradbiomed.2016.02.003
7. Amato SM, Brynildsen MP (2015) Persister heterogeneity arising from a single metabolic stress. Curr Biol 25(16):2090–2098. https://doi.org/10.1016/j.cub.2015.06.034
8. Amato SM, Brynildsen MP (2014) Nutrient transitions are a source of persisters in *Escherichia coli* biofilms. PLoS One 9(3):e93110. https://doi.org/10.1371/journal.pone.0093110
9. Tamae C, Liu A, Kim K, Sitz D, Hong J, Becket E, Bui A, Solaimani P, Tran KP, Yang H, Miller JH (2008) Determination of antibiotic hypersensitivity among 4,000 single-gene-knockout mutants of *Escherichia coli*. J Bacteriol 190(17):5981–5988. https://doi.org/10.1128/JB.01982-07
10. Niba ET, Naka Y, Nagase M, Mori H, Kitakawa M (2007) A genome-wide approach to identify the genes involved in biofilm formation in *E. coli*. DNA Res 14(6):237–246. https://doi.org/10.1093/dnares/dsm024
11. Hansen S, Lewis K, Vulic M (2008) Role of global regulators and nucleotide metabolism in antibiotic tolerance in *Escherichia coli*. Antimicrob Agents Chemother 52 (8):2718–2726. https://doi.org/10.1128/AAC.00144-08
12. Donath MJ 2nd, Dominguez MA, Withers ST 3rd (2011) Development of an automated platform for high-throughput P1-phage transduction of *Escherichia coli*. J Lab Autom 16 (2):141–147. https://doi.org/10.1016/j.jala.2010.08.005
13. Riley M, Abe T, Arnaud MB, Berlyn MK, Blattner FR, Chaudhuri RR, Glasner JD, Horiuchi T, Keseler IM, Kosuge T, Mori H, Perna NT, Plunkett G 3rd, Rudd KE, Serres MH, Thomas GH, Thomson NR, Wishart D, Wanner BL (2006) *Escherichia coli* K-12: a cooperatively developed annotation snapshot—2005. Nucleic Acids Res 34 (1):1–9. https://doi.org/10.1093/nar/gkj405
14. Snyder L, Snyder L (2013) Molecular genetics of bacteria, 4th edn. ASM Press, Washington, DC
15. Thierauf A, Perez G, Maloy AS (2009) Generalized transduction. Methods Mol Biol 501:267–286. https://doi.org/10.1007/978-1-60327-164-6_23
16. Kohanski MA, Dwyer DJ, Hayete B, Lawrence CA, Collins JJ (2007) A common mechanism of cellular death induced by bactericidal antibiotics. Cell 130(5):797–810. https://doi.org/10.1016/j.cell.2007.06.049
17. Zaslaver A, Bren A, Ronen M, Itzkovitz S, Kikoin I, Shavit S, Liebermeister W, Surette MG, Alon U (2006) A comprehensive library of fluorescent transcriptional reporters for *Escherichia coli*. Nat Methods 3(8):623–628. https://doi.org/10.1038/nmeth895

Chapter 10

Genetic Engineering of Cyanobacteria: Design, Implementation, and Characterization of Recombinant *Synechocystis* sp. PCC 6803

Jacob Sebesta, Allison Werner, and Christie Ann Marie Peebles

Abstract

Synechocystis sp. PCC 6803 is a model cyanobacterium which has been investigated to produce a variety of fuels and chemicals. Genetic mutations are of interest for studying photosynthesis and engineering chemical production. Here, methods for culturing, preserving, and genetically transforming *Synechocystis* sp. PCC 6803 are detailed including methods to test promoter strength using the green fluorescent protein reporter. Furthermore, a method for markerless transformation of chromosomal DNA is presented. Sufficient details are provided to enable application by the novice investigator.

Key words Cyanobacteria, *Synechocystis* sp. PCC 6803, Plasmid design, Homologous recombination, Markerless transformation

1 Introduction

Cyanobacteria, formerly called blue-green algae, perform oxygenic photosynthesis using solar light energy and atmospheric carbon dioxide. *Synechocystis* sp. PCC 6803 (hereafter *Synechocystis* 6803) is a model cyanobacterium by virtue of a fully sequenced, publically available genome [1, 2]. *Synechocystis* 6803 performs natural DNA uptake and homologous recombination, making the process of modifying chromosomal DNA relatively easy. Compared to common heterotrophic production chassis such as yeast and *E. coli*, cyanobacteria have a major advantage of not requiring expensive sugar feedstocks (which results in competition with food supply systems) while maintaining high growth rates relative to other photosynthetic organisms such as plants. Many researchers are working to extend metabolic engineering principles developed in

Jacob Sebesta and Allison Werner contributed equally to this chapter.

Christine Nicole S. Santos and Parayil Kumaran Ajikumar (eds.), *Microbial Metabolic Engineering: Methods and Protocols*, Methods in Molecular Biology, vol. 1927, https://doi.org/10.1007/978-1-4939-9142-6_10,

E. coli and yeast to cyanobacteria. Taken together, these advantages have resulted in *Synechocystis* 6803 gaining wide attention as a potential chassis for the production of a variety of chemicals [3]. The objective of this chapter is to describe the protocols necessary for obtaining, culturing, preserving, and genetically modifying *Synechocystis* 6803 with sufficient detail so that new investigators can apply them successfully. An example is provided: constructing and transforming a promoter:*gfp* cassette into the *Synechocystis* 6803 *slr0168* neutral site using markerless transformation and testing promoter strength by measuring fluorescence.

2 Materials

2.1 Cell Culture

1. Glass Erlenmeyer flasks, 250 mL (*see* **Note 1**).
2. Foam stoppers.
3. Glass storage bottles, 1 L.
4. BG-11 Stock Solutions (*see* Table 1), sterile.
5. 3 M hydrochloric acid, sterile.
6. 3 M sodium hydroxide, sterile.
7. Antibiotic solutions (*see* Table 2).
8. Culture plates.
9. Agar.
10. Sodium thiosulfate, sterile.
11. *N*-Tris(hydroxymethyl)methyl-2-aminoethanesulfonic acid, sterile.

Table 1
BG-11 media component recipes

Part Name	Chemical formula and name	Amount in 200 mL
100× Comp. A	$NaNO_3$, Sodium nitrate $MgSO_4{\cdot}7H_2O$, magnesium sulfate heptahydrate $CaCl_2{\cdot}2H_2O$, calcium chloride dehydrate $C_6H_8O_7$, citric acid NH_3FeCitrate, ammonium ferric citrate EDTA IDRANAL	30.00 g 1.498 g 0.720 g 0.120 g 0.120 g 0.020 g
600× Comp. B	$K_2HPO_4{\cdot}3H_2O$, potassium phosphate trihydrate	4.800 g
100× Comp. C	Na_2CO_3, sodium carbonate	0.400 g
1000× Trace Metals	H_3BO_3, boric acid $MnCl_2{\cdot}4H_2O$, manganese II chloride tetrahydrate $ZnSO_4{\cdot}7H_2O$, zinc sulfate heptahydrate $Na_2MoO_4{\cdot}2H_2O$, sodium molybdate dehydrate $CuSO_4{\cdot}5H_2O$, copper II sulfate pentahydrate $Co(NO_3)_2{\cdot}6H_2O$, cobalt II nitrate hexahydrate	0.572 g 0.362 g 0.044 g 0.078 g 0.016 g 0.010 g

Table 2
BG-11 Media Additions

Chemical name	Final Concentration	Stock Concentration	Notes
Kanamycin	50 μg/mL	50 mg/mL	Prepare 1000× stocks by dissolving 500 mg kanamycin salt in 10 mL deionized water. Filter-sterilize using a syringe and 0.2 μm filter into autoclaved 1.5 mL microcentrifuge tubes. Store in 1 mL aliquots at −20 °C
Spectinomycin	50 μg/mL	50 mg/mL	Prepare 1000× stocks by dissolving 500 mg spectinomycin salt in 10 mL deionized water. Filter-sterilize using a syringe and 0.2 μm filter into autoclaved 1.5 mL microcentrifuge tubes. Store in 1 mL aliquots at −20 °C
Sodium thiosulfate	3.3 g/L	1 M	Prepare 300 mL 1 M stock by dissolving 47.43 g sodium thiosulfate in 200 mL using a magnetic stir bar on a stir plate. Once dissolved, adjust volume to 300 mL and filter-sterilize into a sterile bottle
TES[a]	10 mM	1 M	Dissolve 68.78 g TES in 200 mL deionized water. Adjust the pH to 8.0 using NaOH. Adjust volume to 300 mL. Filter-sterilize using a syringe and 0.2 μm filter and store at 4 °C protected from light
DMSO[b]	5%	50%	Filter-sterilize a 50% DMSO in water solution and store at room temperature
Nickel sulfate	20 μM	20 mM	Prepare 1000× stock of 20 mM nickel sulfate by dissolving 0.526 g of nickel sulfate hexahydrate (MW = 262.848) in 100 mL deionized water. Filter-sterilize into an autoclaved glass bottle. Store at room temperature

[a] *N*-Tris(hydroxymethyl)methyl-2-aminoethanesulfonic acid
[b] Dimethyl sulfoxide

12. Glass beads or cell spreaders, sterile.
13. Surgical tape.
14. 50% dimethyl sulfoxide, sterile.
15. Centrifuge tubes, 15–50 mL.
16. Microcentrifuge tubes, 1.5 mL.
17. Gloves.
18. 75% ethanol in spray bottle.
19. Laminar flow hood or biosafety cabinet.
20. Temperature-controlled growth chamber with lights and platform shaker.
21. −80 °C freezer.

2.2 DNA Design for Homologous Recombination

1. Software for visualizing DNA sequences (*see* **Note 2**).

2.3 Plasmid Construction

1. Cloning vectors (*see* Subheading 3.2, **step 3**).
2. Chemically competent *E. coli* strain.
3. DNA polymerase.
4. Nuclease-free water.
5. KpnI restriction enzyme (Thermofisher #ER0521).
6. NdeI restriction enzyme (Thermofisher #ER0581).
7. T4 DNA ligase.

2.4 Natural DNA Uptake and Homologous Recombination

1. Falcon tubes, 14 mL.
2. Cell spreaders, sterilized.
3. Inoculation loops or toothpicks, sterilized.
4. Petri dishes.
5. Glass Erlenmeyer flasks, 250 mL.

2.5 Counterselection for Markerless Transformation

Same materials as Subheading 2.4 *plus* nickel sulfate hexahydrate.

3 Methods

This section describes methods for:

1. Inoculating, maintaining, and preserving *Synechocystis* 6803 cultures.
2. Designing and constructing plasmids for transformation of *Synechocystis* 6803 chromosomal DNA.
3. Transforming the chromosomal DNA of *Synechocystis* 6803 using an antibiotic selection marker.
4. Transforming the chromosomal DNA of *Synechocystis* 6803 using a counter selection marker.

3.1 Cell Culture

All handling of cyanobacteria and media should be performed using sterile technique in a laminar flow hood. Wild-type *Synechocystis* 6803 [N−1] can be obtained from the American Type Culture Collection or the Pasteur Culture Collection (*see* **Note 3**).

1. Prepare BG-11 components as specified in Table 1 by dissolving the solid ingredients in deionized water. Autoclave components A, B, and C on liquid cycle (121 °C, 15 PSIG)

for >30 min. Filter-sterilize the Trace Metals component. Store at 4 °C.

2. To prepare BG-11 liquid media, autoclave 1 L deionized water in a 2 L glass bottle. Using sterile technique in a laminar flow hood, add 10 mL 100× A, 10 mL 600× B (*see* **Note 4**), 10 mL 100× C, and 1 mL 1000× Trace Metals. Adjust the pH to 8.0 using HCl and NaOH as needed. Add 10 mL 1 M TES-NaOH, pH 8.0. Store at 4 °C or room temperature.
3. To prepare BG-11 agar plates (60 mm diameter petri dishes), autoclave 700 mL deionized water with 7 g agar and a stir bar. Allow to cool, while gently stirring. When the media temperature reaches ~55 °C, add 7 mL 100× A, 7 mL 600× B, 7 mL 100× C, and 0.7 mL 1000× Trace Metals. Add 7 mL 1 M TES-NaOH, pH 8.0. In addition, add 14.6 mL 1 M sodium thiosulfate (to 3.3 g/L) and antibiotics as needed (*see* Table 2 and **Note 5**). Pour into plates and let sit until solidified. Agar plates can be stored at 4 °C; however, plates less than 1 week old allow faster cyanobacteria growth.
4. To inoculate and maintain a liquid cyanobacterial culture, add liquid BG-11 and appropriate antibiotics (*see* Table 2) to an Erlenmeyer flask. Inoculate liquid culture by pipetting in cells from another culture (e.g., from freezer stock) or scraping solid culture (e.g., individual colony, agar stab) to 20–30 mL of liquid BG-11. Incubate at 26–30 °C, 225 rpm, 50–200 μmol photos m^{-2} s^{-2} for 2–5 days (*see* **Note 6**).
5. Monitor cell growth by performing cell counts or optical density measurements at 730 nm. Routine culture maintenance should remain below OD_{730} ~2.0. Passage number (the number of times a colony is restreaked or transferred to a new flask) should be kept as low as possible to minimize genetic drift of the strain.
6. To transfer a liquid cyanobacterial culture to solid media, pipette 10–100 μL liquid culture onto a BG-11 agar plate and spread using autoclaved glass beads or cell spreaders (*see* **Note 7**). Seal plate with surgical tape and incubate at 26–30 °C, ~40% humidity, 50–200 μmol photons m^{-2} s^2 for 5–15 days (*see* **Note 8**).
7. To freezer stock a cyanobacterial culture, grow culture in BG-11 supplemented with appropriate antibiotics to OD_{730} 0.6–0.9. Transfer culture to a centrifuge tube and centrifuge at 2760 × *g* for 5 min. Discard supernatant and resuspend cell pellet in 1/10th volume BG-11 liquid media (no antibiotic). Add DMSO (*see* Table 2) to final concentration of 5%. Aliquot into 1.5 mL microcentrifuge tubes and store at −80 °C.

3.2 Plasmid Design and Construction

1. Retrieve the DNA sequence of interest. Native *Synechocystis* 6803 genes can be accessed from Cyanobase (http://genome.microbedb.jp/cyanobase/), and nonnative genes can be found on NCBI GenBank or an organism-specific database (*see* **Note 9**). Save the sequence as a new file in your choice of DNA editing software. If a genetic knockout is the end goal, the gene of interest should be an antibiotic resistance cassette or other selectable marker.
2. Determine which expression control elements will be applied. Promoters and ribosome bindings sites (RBSs) are important control elements to consider for strain design. Several native *Synechocystis* 6803 promoters, as well as modified *E. coli* promoters that perform in *Synechocystis* 6803, have been previously characterized [4–7]. The RBS Calculator is a valuable tool for assessing native RBSs and designing sequences for strong translation [8]. However, the accuracy of the RBS Calculator in predicting translation initiation rates in Synechocystis 6803 is still being investigated [9]. Codon optimization is frequently employed to optimize heterologous gene expression, especially with genes derived from eukaryotes. Commercial gene synthesis companies typically offer codon optimization free of charge when synthetic genes are ordered. However, codon optimization is not fully understood, and the wide variety of mechanisms by which codon usage impacts gene expression introduces a degree of uncertainty to optimization. A good review of codon usage in this context has been provided by Quax and colleagues [10].
3. Selection of a cloning vector to use for assembling the DNA construct in *E. coli* should be informed by the vector copy number, selection marker, multiple cloning site (MCS), and presence of homologous regions to *Synechocystis 6803* neutral sites (*see* **Note 10**). Several plasmids have been developed recently that can replicate in both *E. coli* to facilitate cloning, as well as in *Synechocystis* 6803 for testing the plasmid [11]. For example, pIGA4 is a plasmid used in our lab to test promoter strengths. It uses the pUC origin of replication to achieve a high copy number in *E. coli*, contains the ampicillin antibiotic resistance marker, and homologous regions (HRs) for the transformation into the *slr0168* neutral site [4]. Between the homologous regions, the plasmid contains the gene sequence for a green fluorescent protein variant mutated to degrade more quickly than the wild type, facilitating measurement of rapid changes in expression [4]. The plasmid backbone has been mutated to remove an NdeI cut site from the backbone so that KpnI and NdeI cut sites can be used to clone in and test different promoters. If a vector is used that does not

have HRs, they need to be introduced flanking the gene of interest (or MCS) and selection marker.

4. Select a cloning method to insert the gene of interest into the cloning vector. Many options exist, including "cut-and-paste" using restriction enzymes and ligase, GoldenGate [12], ligase cycling [13], Gibson assembly [14], and others. In the simplest cases, a gene can be inserted into the MCS of a vector using a cut-and-paste approach, explained here.
5. Design PCR primers to copy the gene or DNA region of interest (*see* **Note 2** for DNA editing software options). In general, primers should be 15–30 bases long, have a melting temperature (T_m) between 50 and 60 °C, finish with a G or C at the 3′ end, and have minimal predicted primer dimers or secondary structure. Additionally, the forward and reverse primers melting temperatures should be within 5 °C of each other. If using cut-and-paste to clone into the MCS, add restriction endonuclease recognition sequences to 5′ end of both primers. In this case, four bases should be added on the far 5′ end to facilitate enzyme binding.
6. Follow standard cloning practices for the method of choice to assemble the plasmid.

3.3 Natural DNA Uptake and Homologous Recombination

Synechocystis 6803 is naturally competent (i.e., naturally uptakes DNA) due to the presence of a type IV-like pilus structure [15]. Outlined here is a protocol for transforming *Synechocystis* 6803. This protocol takes into consideration some of the findings of Zang et al., which previously examined the effects of variables such as homologous region length and the time of incubation with plasmid DNA on transformation efficiency [16]. Though it is not included in this protocol, Wang et al. showed that premethylation of the plasmid DNA used to transform by methyltransferase C from *Synechocystis* 6803 improved transformation efficiency [17].

1. To concentrate *Synechocystis* 6803 cells for natural DNA uptake, grow cells to mid-log phase ($OD_{730} = 0.5–0.75$). For each transformation, aliquot 10 mL of cell culture into a 15 mL conical centrifuge tube on ice. Centrifuge each tube at 4000 × *g* for 6 min at 4 °C. In addition to the tubes destined for transformation, prepare a no-DNA negative control. Discard the supernatant, resuspend the cell pellet in 100 μL of BG-11, and transfer suspension to a 14 mL round-bottom falcon tube.
2. Add 1 μg of the plasmid being transformed to the suspension and incubate cells for 5 h under low light (20–100 μmol photons m^{-2} s^{-1}) at 30 °C. Tubes should be gently shaken by hand after ~2.5 h to resuspend the cells.

3. Pipette 100 μL of the transformation mixture onto a BG-11 agar plate without antibiotics and spread using a cell spreader. Seal the plate with surgical tape and incubate under low light at 30 °C for 48 h for recovery. The plate should be covered with a dense green film of cells at this point.
4. After 48 h, select for transformants (colonies that have the desired mutation). Use a bacterial loop spreader to perform a triple streak procedure from the recovery plate to a BG-11 plate supplemented with antibiotics (*see* Table 2). To isolate individual colonies, perform a triple streak by touching the loop spreader to the plate and tracing a long, serpentine path through one third of a fresh plate containing antibiotics. Flame the spreader to sterilize. Once the spreader is cool, the loop is streaked through part of the first serpentine path to gather some cells but less than was initially gathered from the plate lacking antibiotics. These cells are then spread through a second third of the plate in a second serpentine path. The loop is flamed and cooled again, then swept through the second serpentine path to collect cells and streaked in the last third of the plate in a third serpentine path. Seal the plate with surgical tape and incubate under low light at 30 °C. Single colonies should be observed in 5–14 days.
5. To verify that your gene of interest has been inserted by homologous recombination, perform a colony PCR (cPCR) with primers that bind-to and amplify the region of modification. Briefly, design primers which bind to the homologous region outside of the DNA inserted into the cloning vector (*see* Subheading 3.2, **step 5** for primer design). As a negative control, perform a cPCR on wild-type *Synechocystis* 6803 (*see* **Note 11**). Since *Synechocystis* 6803 maintains multiple copies of its chromosome, some copies may have the wild-type sequence, while others have the desired mutation.
6. To achieve full segregation, restreak single colonies on a fresh BG-11 plate with antibiotics and place in 30 °C lighted incubator. Repeat this process once a week for 3–4 weeks or until full segregation is achieved (*see* **Note 12**). Sequence confirmation of segregated mutants is highly recommended.

3.4 Counterselection for Markerless Recombination

To remove selection markers, counterselection methods have been developed for transformation of *Synechocystis* 6803. This allows for the accumulation of multiple mutations without the potential burden of expressing multiple antibiotic resistance genes. Described here is a counterselection method that first inserts both nickel-inducible expression of *mazF*, a toxic endoribonuclease, and *aphII*, kanamycin resistance, into the locus of interest [18]. Nickel induction of the *mazF* gene is due to the use of the native, nickel-responsive *nrsB* promoter.

To perform markerless recombination, two transformations are necessary: (1) insert the *mazF/kanR* cassette into the HR of interest, and (2) replace *mazF/kanR* with the modification/gene of interest. The following protocol describes the steps necessary to select against cells retaining the *mazF/aphII* cassette and for the cells that obtained the modification of interest (*see* **Note 13**). Here we assume that the starting strain already contains the *mazF/KanR* cassette inserted into the locus of interest (*see* Subheading 3.3 for transformation technique) and that a new plasmid has been constructed to replace the cassette using the same homologous regions (*see* Subheading 3.2 for plasmid design and construction).

1. To prepare *Synechocystis* 6803 for natural DNA uptake, follow **steps 1** and **2** in Subheading 3.3, above.
2. Following the 5 h incubation with the plasmid DNA, centrifuge the cells at 4000 × *g* for 6 min. Discard the supernatant and resuspend the cell pellet in a small amount of fresh BG-11 (100–500 μL).
3. Add the resuspended cells to 10 mL of fresh BG-11 in a shake flask and incubate for an additional 24 h.
4. With a cell spreader, spread 100 μL of each culture on BG-11 agar plates supplemented with 20 μM nickel sulfate (*see* Table 2). Seal the sides of the plates with surgical tape. Incubate the plates under low light at 30 °C, 40–50% humidity, for 5–7 days.
5. Once colonies form, streak single colonies on both BG-11 plus nickel plates and BG11 plus kanamycin plates. The colonies with the expected insert should survive on BG-11 plus nickel but not on BG-11 plus kanamycin plates. Colonies may need to be restreaked two or more times on nickel plates before the chromosomal copies are completely segregated (*see* **Note 14**).
6. Colony PCR (cPCR) may be used to confirm the presence of the mutations as well as the absence of the wild-type fragment. Primers that anneal to opposite strands of either homologous region pointing into the insert may be used if the size of the insert has a significantly different length than the *mazF/kanR* cassette. If the cPCR results in multiple bands, both versions of the chromosome are present. Alternatively, two reactions for each colony may provide better evidence for presence/absence of the two versions. One reaction would check for the presence of the correct mutation by including primers within the homologous region and the gene being inserted, while the second would confirm the absence of the *mazF/kanR* cassette by including primers within the homologous region and within the *mazF/kanR* cassette. For such dual reaction tests, full segregation is indicated when only the first of these reactions produces a band.

7. To start a liquid culture, take a swab of cells from the BG-11 plus nickel plate and place it in 1 mL of BG-11 liquid media supplemented with 20 μM nickel sulfate in a 14 mL round-bottom Falcon tube. Once the mutation has been confirmed, the addition of nickel sulfate is no longer necessary. Incubate at 30 °C with shaking. Once cells reach a sufficient cell density (about 2 days), inoculate 50 mL of BG-11 + 20 μM nickel with the cell suspension. The culture should be transferred to fresh media every 6–7 days.

4 Example: Testing Promoter Strength in *Synechocystis* 6803

In this example, step-by-step instructions will be provided for creating a strain of *Synechocystis* 6803 which contains a promoter driving green fluorescent protein (GFP) expression within the *slr0168* neutral site via markerless homologous recombination. A flow-diagram of these methods is provided (*see* Fig. 1).

4.1 Obtain Parent Plasmids

Two plasmid backbones are needed for this protocol: pMK0168 and pIGA4 (Addgene). pMK0168 contains a nickel-inducible *mazF* and kanamycin resistance marker targeted to the *slr0168* neutral site (*see* Fig. 1a). pIGA4 contains the GFP marker and flanking *slr0168* homologous regions; the promoter of interest will be inserted into the MCS immediately upstream of GFP (*see* Fig. 1a).

4.2 Amplify Promoter of Interest

Design primers to amplify the promoter of interest. The forward primer should be designed to contain a KpnI recognition site (GGTACC) and the reverse primer should be designed to contain a NdeI recognition site (CATATG). Perform PCR on the promoter of interest and purify the amplified promoter fragment. The resulting fragment will contain KpnI and NdeI restriction enzyme cut sites (*see* Fig. 1b). For example, to clone the promoter from principal RNA polymerase sigma factor P*sigA*, use forward primer (5′-TATAAGGTACCCCGTCATCGATTC-3′) and reverse primer (5′-TATAACATATGGCCGTTTTCCTCGTTAAC-3′) where the underlined bases denote restriction enzyme recognition sites.

4.3 Construct Promoter:GFP Plasmid

Digest both pIGA4 and the promoter fragment with KpnI and NdeI at 37 °C for 20 min to 1 h followed by heat inactivation at 80 °C for 20 min. Dephosphorylate the sticky ends of pIGA4 using a phosphatase such as shrimp alkaline phosphatase or calf intestinal phosphatase; incubate at 37 °C for 20 min to 1 h. Purify all products using a PCR purification kit. Ligate the digested promoter into pIGA4 using T4 DNA ligase overnight at 16 °C. Transform the ligation product into chemically competent DH5α *E. coli* via a heat shock protocol and allow the cells to recover in LB media without antibiotic for 1 hour. Plate transformed *E. coli* onto LB agar plates supplemented with ampicillin and incubate at 37 °C for ~16 h. Perform colony PCR on colonies and extract the plasmid from correct colonies. Confirm the plasmid is constructed correctly with sequencing (*see* Fig. 1c).

4.4 Transform Synechocystis 6803 with Counterselection Plasmid pMK0168

Follow the steps described in Subheading 3.3 to insert *mazF/kanR* cassette into the *slr0168* neutral site (*see* Fig. 1d). Perform colony PCR (cPCR) of *Synechocystis* colonies to confirm proper transformation using the primers listed in Table 3. The same protocols used to complete cPCR of *E. coli* may be used for *Synechocystis*. Figure 1f shows the binding sites of the primers used for cPCR. If *mazF/kanR* is properly inserted, cPCR with primers #3 and #4 will produce a 4.4 kb band; primers #3 and #5 will produce a 4.2 kb band. If proper insertion was not achieved, Primers #3 and #4 should produce a 0.6 kb band, and no bands should be produced from primers #3 and #5. Incomplete segregation would be indicated if primers #3 and #4 produce both a 4.4 kb band and a 0.6 kb band.

4.5 Synechocystis 6803 Transformation with Promoter:GFP Plasmid

Follow the steps described in Subheading 3.4. To replace the *mazF/kanR* cassette with the gene of interest in the *slr0168* neutral site (*see* Fig. 1e). Transformants should grow on BG-11 + Ni agar plates but not on BG-11 + Kan agar plates.

Ensuring complete segregation of the transformed cyanobacteria strain is essential. To do this, perform two PCR reactions for each colony using the primers listed in Table 3. If the GFP expression cassette is properly inserted, cPCR with primers #3 and #4 will produce a 2.6 kb band and primers #3 and #6 will produce a 2.4 kb band. If the mazF/kanR cassette remains in the slr0168 site, primers #3 and #4 will produce a 4.4 kb band and primers #3 and #6 will produce no band (*see* **Note 15**). Sequence verify either the PCR fragment produced from primers #3 and #4, or the genomic DNA, using primer #3.

4.6 Test Promoter by Measuring Fluorescence

Cultivate cells using the steps described in Subheading 3.1. Inoculate wild-type and transformed cultures to OD_{730} ~0.1 from seed cultures at OD_{730} ~0.6–1.0 in triplicate. To measure fluorescence, transfer 300 μL of culture to a black 96-well flat-bottom plate. Using a plate reader with fluorescence capabilities, measure fluorescence with excitation wavelength 485 nm and emission wavelength of 525 nm. Measure optical density and fluorescence every 6–12 h for as many days as desired. Relative fluorescence will provide a measure of promoter strength and activity over the test period.

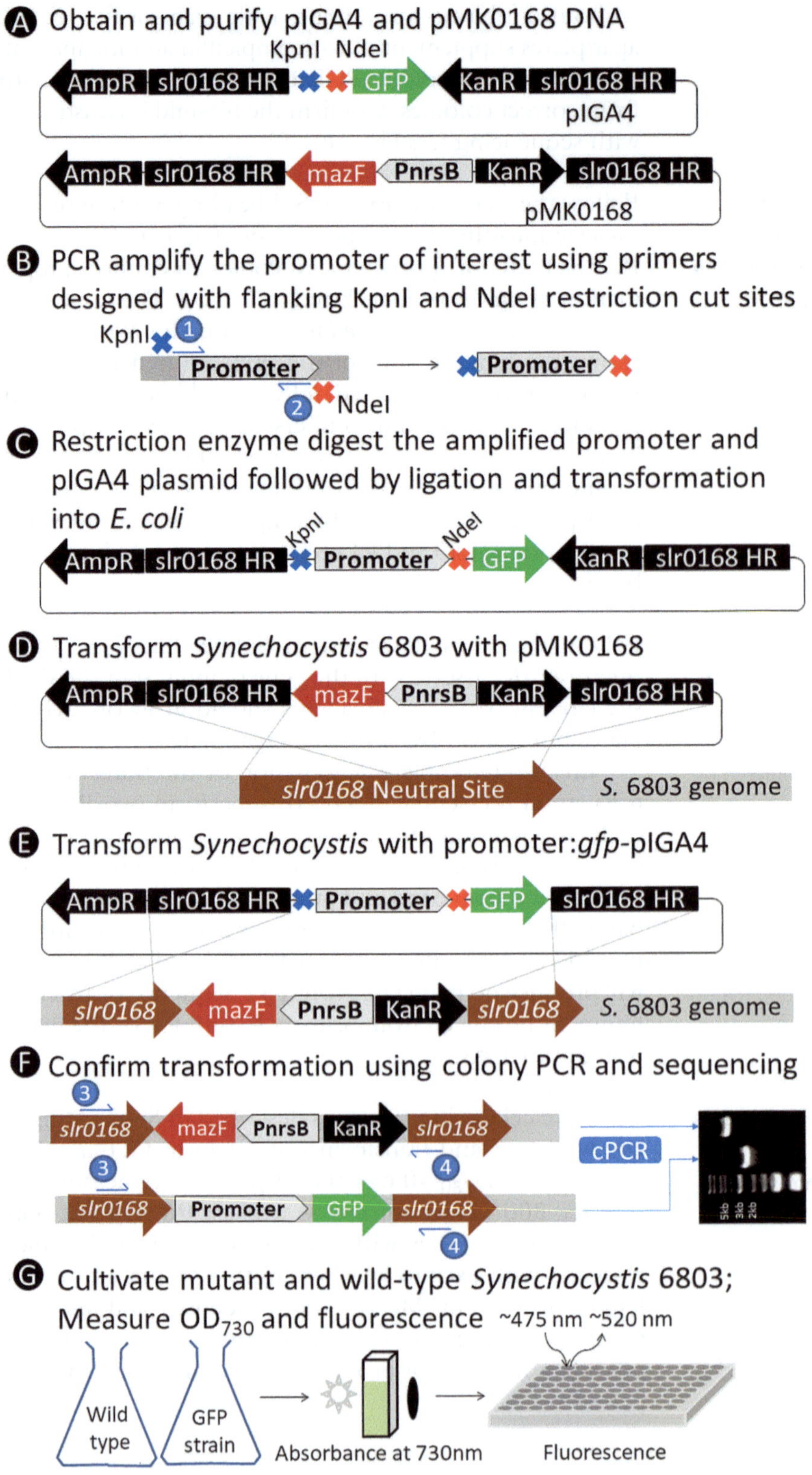

Fig. 1 Design, implementation, and characterization of *Synechocystis* 6803 transformants with a promoter of interest driving GFP production. Panel A shows the promoter characterization plasmid and counterselection

Table 3
Primers

Primer #	Primer name	5′ → 3′ sequence
1	PsigA forward	5′-TATAAGGTACCCCGTCATCGATTC-3′
2	PsigA reverse	5′-TATAACATATGGCCGTTTTCCTCGTTAAC-3′
3	Slr0168 up forward	5′-TCCAGGCCACATTGTTG-3′
4	Slr0168 dn reverse	5′-AGTGACCTATTCAATCAGGAAGG-3′
5	MK reverse	5′-CAGAAAGACTTAAAATTATTGCCG-3′
6	Gfp reverse	5′-TTATTATTTGTATAGTTCATCCATGCC-3′

5 Notes

1. We recommend an acid wash of flasks after culturing *Synechocystis* 6803.
2. APlasmidEditor (Ape, http://biologylabs.utah.edu/jorgensen/wayned/ape/) is a free tool for visualizing DNA sequences. SnapGene (http://www.snapgene.com/) can produce better plasmid images and may be more intuitive to use, but requires a license to use. IDT Oligo Analyzer (https://www.idtdna.com/calc/analyzer) is a tool and service for analyzing and ordering primers.
3. Several strains of *Synechocystis* 6803 are available for purchase and have different characteristics, including the ability to grow with light-activated heterotrophic growth. The strain of cyanobacteria referred to here (ATCC® 27,184™) can be purchased here: https://www.atcc.org/Products/All/27184.aspx. *Synechocystis* 6803 is also available through the Pasteur Culture Collection at https://brclims.pasteur.fr/crbip_catalogue/faces/recherche_catalogue.xhtml.
4. Since BG-11 has been previously found to be limiting in phosphate [19], Component B can be prepared as a 600× solution, but still added to the BG-11 media preparation at a 100× dilution.
5. Although chloramphenicol is reported as an appropriate antibiotic for selection in *Synechocystis* 6803 and other cyanobacteria, in our hands, selection is poor. We recommend

Fig. 1 (continued) plasmid needed for this protocol, and Panel B, the promoter PCR. Panel C shows the constructed promoter characterization plasmid. Panels D and E show the two transformation steps, while Panel F shows the colony PCR used to screen for correct transformants, and Panel G shows the characterization of the promoter:gfp construct in *Synechocystis* 6803

Kanamycin or Spectinomycin if given a choice of antibiotic resistance.

6. Commercially available fluorescent lights are commonly used. LED lights may also be used. "Grow lights" targeting specific wavelengths should not be used as these trigger the reorganization of pigment–protein complexes in *Synechocystis* 6803 which may alter growth.
7. If the culture is sufficiently dense, adding small volumes (e.g., 10 μL) will be difficult to spread. Add liquid BG-11 up to 100 μL (e.g., 90 μL) on plate before spreading.
8. Using surgical tape, as opposed to Parafilm®, is advantageous for incubation of BG-11 plates because it allows for more efficient gas transfer into the plate and thus faster cell growth.
9. The reader is referred to the Eaton-Rye methods paper for details on how to locate and download sequences from Cyanobase [20]. *Synechocystis* 6803 sequences can be accessed at Cyanobase: http://genome.microbedb.jp/cyanobase/Synechocystis. Gene sequences from many organisms can be obtained at NCBI GenBank: https://www.ncbi.nlm.nih.gov/genbank/.
10. Several neutral sites have been identified in *Synechocystis* 6803 [21].
11. If expected PCR product is similar to the negative control, then a second primer specific to the DNA insert can be used.
12. *Synechocystis* 6803 often maintains multiple copies of its chromosome. Full segregation means that the genetic modification of interested was achieved on each copy of the chromosome. Interestingly, the first round of transformation rarely results in full segregation, but it can be achieved by repeatedly restreaking the same colony onto selective media.
13. To perform markerless recombination as these protocols are written, the investigator should apply the protocols in Subheading 3.3 using nickel-inducible mazF/KanR cassette as the gene interest. To select for the successful transformants, BG-11 agar plates supplemented with only kanamycin should be used. Then, this base strain should be applied for further modification using the protocols in Subheading 3.4 using the gene of interest.
14. In our laboratory, we have observed the formation of both small and large colonies. To increase the probability of observing segregated transformants, screen a mixture of both large and small colonies.
15. Colonies may be screened for the desired mutation as soon as they appear. The presence of multiple copies of the chromosome in *Synechocystis* 6803 complicates screening for

transformants as some copies may carry the mutation while others retain the wild-type sequence. Fully segregated transformants are those that carry the mutation and have completely lost the wild-type sequence (or the mazF/kanR cassette in the case of a markerless selection transformation) between the homologous regions used. Screening colonies using colony PCR should be conducted with primers that bind within the homologous regions in order to capture both the mutated and wild-type sequences (if both are present). Alternatively, two PCR reactions may be completed for each colony, one with primers that amplify only the mutated sequence, and a second with primers that amplify only the wild-type sequence.

Acknowledgments

The authors acknowledge funding for this work by the National Science Foundation (NSF awards 1336236 and 1332404). Stevan Albers and Yi Ern Cheah contributed to this work by developing the protocols for transforming *Synechocystis 6803* used in our lab and teaching us these methods.

References

1. Kaneko T, Sato S, Kotani H et al (1996) Sequence analysis of the genome of the unicellular Cyanobacterium *Synechocystis* sp. strain PCC6803. II. Sequence determination of the entire genome and assignment of potential protein-coding regions. DNA Res 3:109–136. https://doi.org/10.1093/dnares/3.3.109
2. Nakao M, Okamoto S, Kohara M et al (2010) CyanoBase: the cyanobacteria genome database update 2010. Nucleic Acids Res 38: D379–D381. https://doi.org/10.1093/nar/gkp915
3. Angermayr SA, Gorchs Rovira A, Hellingwerf KJ (2015) Metabolic engineering of cyanobacteria for the synthesis of commodity products. Trends Biotechnol 33:352–361. https://doi.org/10.1016/j.tibtech.2015.03.009
4. Albers SC, Gallegos VA, Peebles CAM (2015) Engineering of genetic control tools in *Synechocystis* sp. PCC 6803 using rational design techniques. J Biotechnol 216:36–46. https://doi.org/10.1016/j.jbiotec.2015.09.042
5. Huang HH, Camsund D, Lindblad P, Heidorn T (2010) Design and characterization of molecular tools for a synthetic biology approach towards developing cyanobacterial biotechnology. Nucleic Acids Res 38 (8):2577–2593. https://doi.org/10.1093/nar/gkq164
6. Camsund D, Lindblad P (2014) Engineered transcriptional systems for cyanobacterial biotechnology. Front Bioeng Biotechnol 2:40
7. Albers SC, Peebles CAM (2016) Evaluating light-induced promoters for the control of heterologous gene expression in *Synechocystis* sp. PCC 6803. Biotechnol Prog 33(1):45–53. https://doi.org/10.1002/btpr.2396
8. Salis HM (2011) Chapter two - the ribosome binding site calculator. In: Enzymology CVBT-M in (ed) synth. Biol. Part BComputer aided des. DNA Assem. Academic Press, Cambridge, pp 19–42
9. Wang B, Eckert C, Maness P-C, Yu J (2018) A genetic toolbox for modulating the expression of heterologous genes in the cyanobacterium *Synechocystis* sp. PCC 6803. ACS Synth Biol 7:276–286. https://doi.org/10.1021/acssynbio.7b00297
10. Quax TEF, Claassens NJ, Söll D, van der Oost J (2015) Codon bias as a means to fine-tune gene expression. Mol Cell 59:149–161. https://doi.org/10.1016/j.molcel.2015.05.035
11. Liu D, Pakrasi HB (2018) Exploring native genetic elements as plug-in tools for synthetic biology in the cyanobacterium Synechocystis sp. PCC 6803. Microb Cell Fact 17. https://doi.org/10.1186/s12934-018-0897-8

12. Engler C, Marillonnet S (2014) Golden Gate Cloning. In: Valla S, Lale R (eds) DNA cloning Assem. Methods. Humana Press, Totowa, NJ, pp 119–131
13. de Kok S, Stanton LH, Slaby T et al (2014) Rapid and reliable DNA assembly via ligase cycling reaction. ACS Synth Biol 3:97–106. https://doi.org/10.1021/sb4001992
14. Gibson DG, Young L, Chuang R-Y et al (2009) Enzymatic assembly of DNA molecules up to several hundred kilobases. Nat Methods 6:343–345
15. Yoshihara S, Geng X, Okamoto S et al (2001) Mutational analysis of genes involved in Pilus structure, Motilityand transformation competency in the unicellular motile Cyanobacterium *Synechocystis* sp. PCC6803. Plant Cell Physiol 42:63–73. https://doi.org/10.1093/pcp/pce007
16. Zang X, Liu B, Liu S et al (2007) Optimum conditions for transformation of *Synechocystis* sp. PCC 6803. J Microbiol 45:241–245
17. Wang B, Yu J, Zhang W, Meldrum DR (2015) Premethylation of foreign DNA improves integrative transformation efficiency in *Synechocystis* sp. strain PCC 6803. Appl Environ Microbiol 81(24):8500–8506. https://doi.org/10.1128/AEM.02575-15
18. Cheah YE, Albers SC, Peebles CAM (2013) A novel counter-selection method for markerless genetic modification in *Synechocystis* sp. PCC 6803. Biotechnol Prog 29:23–30. https://doi.org/10.1002/btpr.1661
19. Kim HW, Vannela R, Zhou C et al (2010) Photoautotrophic nutrient utilization and limitation during semi-continuous growth of *Synechocystis* sp. PCC6803. Biotechnol Bioeng 106:553–563. https://doi.org/10.1002/bit.22724
20. Eaton-Rye JJ (2011) Construction of gene interruptions and gene deletions in the Cyanobacterium *Synechocystis* sp. strain PCC 6803. In: Carpentier R (ed) Photosynth. Res. Protoc. Humana Press, Totowa, NJ, pp 295–312
21. Pinto F, Pacheco CC, Oliveira P et al (2015) Improving a *Synechocystis*-based photoautotrophic chassis through systematic genome mapping and validation of neutral sites. DNA Res 22(6):425–437. https://doi.org/10.1093/dnares/dsv024

Chapter 11

Genetic Tools for Streamlined and Accelerated Pathway Engineering in *Yarrowia lipolytica*

Lynn Wong, Benjamin Holdridge, Jake Engel, and Peng Xu

Abstract

Yarrowia lipolytica is an industrial oleaginous yeast that has many attractive physiological and metabolic characteristics for various biotechnological applications. Although it has a long history of industrial applications, the number of genetic tools available to effectively and efficiently engineer *Y. lipolytica* still falls behind the vast number of tools available for common organisms such as *Escherichia coli* and *Saccharomyces cerevisiae*. In this protocol, we have developed a complete and versatile genetic toolkit tailored for facile genetic manipulation in *Y. lipolytica*. We created a versatile DNA assembly platform YaliBrick, which can streamline the cloning of large multigene pathways with reused genetic parts. We established a sensitive luciferase reporter assay to characterize a set of 12 native promoters. In addition, we used YaliBrick to generate different gene configurations in multigene constructs. The five-gene biosynthetic pathway of the anticancer, antimicrobial pigment violacein was rapidly assembled in 1 week to demonstrate the simplicity and effectiveness of integrating pathway-balancing strategies with our YaliBrick vectors. In the end, we incorporated CRISPR-Cas9 into our YaliBrick vectors and achieved indel mutation and frameshift gene deletion at the CAN1 (arginine permease) genomic loci of *Yarrowia lipolytica*. The reported protocol provides a standard procedure to streamline and accelerate metabolic pathway engineering in *Yarrowia lipolytica*.

Key words YaliBrick, Gene assembly, Reporter, Combinatorial pathway library, Promoter engineering, Metabolic engineering, Genome editing, Synthetic biology

1 Introduction

The oleaginous yeast *Yarrowia lipolytica* has become the organism of choice for the production of oleochemicals [1, 2], biofuels [3, 4], and acetyl CoA-derived metabolites [5, 6]. It has been extensively studied as a model organism for lipid accumulation [7] and degradation [8], dimorphism [9, 10], and protein secretory pathway [11]. *Y. lipolytica* is also recognized as a generally regarded as safe (GRAS) organism [12] for the production of large quantities of citric acid [13, 14], α-ketoglutarate [15], and succinic acid [16] in the food industry. Its ability to natively utilize a wide range of

Christine Nicole S. Santos and Parayil Kumaran Ajikumar (eds.), *Microbial Metabolic Engineering: Methods and Protocols*, Methods in Molecular Biology, vol. 1927, https://doi.org/10.1007/978-1-4939-9142-6_11,

substrates, such as glucose, fructose, glycerol, and hydrocarbons, and its low pH tolerance and strictly aerobic nature [2, 6] make this yeast a very attractive candidate for industrial biotechnology applications. Notable examples of its industrial applications include the production of eicosapentaenoic acid (EPA)-rich products [17, 18] and conjugated linoleic acid [19], and cost-efficient production of lipid biofuels [4, 20, 21].

Recent advances in understanding of *Y. lipolytica*'s metabolic pathways, the sequencing of its genome [22] and further expansion of its genetic toolbox have resulted in several successful endeavors to metabolically engineer this unconventional yeast to produce high-value compounds (e.g., limonene [23] and γ-decalactone [24, 25]). These achievements demonstrate the potential of *Y. lipolytica* as an industrial microbe to produce novel compounds and expand beyond its regular portfolio of fatty acids, fatty alcohols, biofuels, and protein production.

Although a number of gene overexpression and deletion tools have been established in *Y. lipolytica* [26], the library of available tools is not as developed as that of other yeasts such as *Saccharomyces cerevisiae*. This is due in part to *Y. lipolytica* having a high frequency of nonhomologous end-joining (NHEJ), while *S. cerevisiae* mainly utilizes its highly efficient homologous recombination repair machinery to repair DNA breaks [27, 28]. Fickers et al. developed a gene disruption method by combining the sticky-end polymerase chain reaction (SEP) method and *Cre-lox* recombination system [29]. However, acceptable rates of recombination only occur when long homology regions (0.5–1 kb) are used, making this method less efficient in the absence of the *KU70* gene deletion [27]. Another hurdle in efficient genome-editing in *Y. lipolytica* lies in the fact that it has a less-defined noncoding RNA processing system. To solve this challenge, Schwartz et al. [30] has used a hybrid tRNA promoter to express gRNA, thus allowing for efficient CRISPR-Cas9 targeting in *Yarrowia*. Alternatively, Gao et al. [31] demonstrated that the Type II RNA polymerase, in combination with the highly efficient hammerhead ribozyme and HDV ribozyme, could efficiently deliver gRNA and achieved multiplex genome-editing in *Y. lipolytica*.

In addition to genome-editing tools, advances have been made to develop a multicopy plasmid-based gene expression system, which is suitable for gene-expression fine-tuning due to the simplicity of plasmid construction, high transformation efficiency, and rapid turnover compared to the laborious and time-consuming genome manipulation process [27, 32]. For example, Liu et al. engineered the low-copy CEN plasmid by fusing different promoters upstream of the centromeric region to improve copy number and gene expression [32].

In light of these advances, there is a pressing need for further development of facile genetic tools that are tailored for modular and combinatorial pathway engineering in *Y. lipolytica* [33, 34]. Effective metabolic engineering of microorganisms relies on assembling and rewiring biosynthetic pathways to channel flux into producing compounds of interest [35–37]. One of the most important aspects is to modulate gene expression of crucial genes to ideally achieve a balanced pathway with no metabolic bottlenecks [38–40]. As such, we have engineered the highly modular vector cloning system YaliBrick to allow for rapid assembly of multigene pathways with high customizability of genetic control elements (promoters, intronic sequences, and terminators). We have also developed a stable, consistent, and reliable reporter system using NanoLuc luciferase to monitor gene expression and serve as a characterization tool for different gene expression control elements.

In this protocol, we will describe the procedure of using the highly sensitive bioluminescence reporter NanoLuc luciferase in *Y. lipolytica* to characterize 12 endogenous promoters that have broad range of transcriptional dynamics. We will also demonstrate the versatility of our YaliBrick vectors to construct gene clusters with unique transcriptional configurations: operon, pseudo-operon, monocistronic. In addition, we constructed a 12 kb, 5-gene biosynthetic pathway for the antibiotic pigment violacein using YaliBrick as a proof of concept for quick pathway assembly and screening. We incorporated our YaliBrick vector with genome-editing feature and successfully introduced indel mutation and frameshift deletion of the CAN1 (arginine permease) genomic loci of *Yarrowia lipolytica*. The YaliBrick vector system will provide a powerful platform to facilitate modular manipulation of multi-gene pathways and the construction of complex genetic circuits in *Y. lipolytica*. All the YaliBrick vectors will be deposited to Addgene to promote collaboration and dissemination of the reported work.

2 Materials

2.1 Reagents and Molecular Kits

Molecular Biology Agarose, Sigma-Aldrich, Cat. No. A6013-250G.

1 kb Plus DNA Ladder, Invitrogen, Cat. No. 10787018.

Ampicillin Sodium Salt, Sigma-Aldrich, Cat. No. A9518-25G.

S.O.C. media, Invitrogen, Cat. No. 15544-034.

10 mg/mL ethidium bromide and 50× TAE buffer, Bio-Rad.

LB Miller base, BD Bioscience, Cat. No. 241420.

Agar, BD Bioscience, Cat. No. 214010.

CSM-Leu, Sunrise Science, Cat. No. 1005-100.

Yeast Nitrogen Base w/o amino acids, Amresco, Cat. No. J386-500G.

Zymolyase®-20 T 1 g, Sunrise Science, Cat. No. N0766391.

D-(+)-Glucose, Sigma-Aldrich, Cat. No. G7021-1KG.

Peptone, BD Bioscience, Cat. No. 211677.

Yeast extract, BD Bioscience, Cat. No. 212750.

Q5® DNA polymerase and buffer, New England Biolabs, Cat. No. M0491L.

T4 DNA ligase and buffer, New England Biolabs, Cat. No. M0202L.

NEB Gibson Assembly Cloning Kit, New England Biolabs, Cat. No. E5510S.

NEB5α high efficiency chemically competent cell, New England Biolabs, Cat. No. C2987H.

Fermentas Fast Digest Enzymes *Dpn*I, *Hind*III, *Eco*RI, *Kpn*I, *Sna*BI, *Xho*I, *Avr*II, *Xba*I, *Spe*I, *Nhe*I, *Sal*I, and *Cla*I.

Zyppy™ Plasmid Miniprep Kit, Zymo Research, Cat. No. D4037.

Zymoclean™ Gel DNA Recovery Kit, Zymo Research, Cat. No. D4001.

DNA Clean & Concentrator™-25, Zymo Research, Cat. No. D4005.

PureLink™ Genomic DNA Mini Kit, Invitrogen, Cat. No. K1820-02.

Wizard Genomic DNA Extraction kit, Promega, Cat. No. A1125.

GoTaq® G2 Hot Start Polymerase, Promega, M7405.

Nano-Glo® Luciferase Assay System, Promega, Cat. No. N1110.

2.2 Strains, Plasmids, and Primers

Please refer to Tables 1 and 2 for strains, plasmids, and primers used in this protocol.

2.3 Equipment

Eppendorf Research Plus pipette, 2.5 μL, 20 μL, 200 μL, and 1000 μL.

Eppendorf Benchtop Microcentrifuge 5430.

Bio-Rad ChemiDOC™ XRS Molecular Imager.

Fisher Scientific Isotemp Multi-block heater.

Sartorius CERTOMAT® BS-1 shaker incubator.

Eppendorf I-26 and E-25 shaker incubator.

Fisher Scientific Isotemp Incubator.

New Brunswick ultralow temperature freezer.

Mettler Toledo balance.

Table 1
Strains and plasmids in this protocol

Plasmid or strain	Relevant properties or genotype	Source
Plasmids		
pYaliA1-EcBirA	pYaliA1 carrying *E. coli* biotin ligase	This study
pYaliA1-VioDCBAE	pYaliA1 carrying *vioD*, *vioC*, *vioB*, *vioA*, *vioE* from *C. violaceum*	This study
pYaliAl-NLuc	pYaliA1 carrying NanoLuc luciferase (TEF promoter)	This study
pYaliB1-NLuc	pYaliB1 carrying NanoLuc luciferase (DGA1 promoter)	This study
pYaliC1-NLuc	pYaliC1 carrying NanoLuc luciferase (ACC promoter)	This study
pYaliD1-NLuc	pYaliD1 carrying NanoLuc luciferase (ACL2 promoter)	This study
pYaliE1-NLuc	pYaliE1 carrying NanoLuc luciferase (IDH2 promoter)	This study
pYaliF1-NLuc	pYaliF1 carrying NanoLuc luciferase (FAS2 promoter)	This study
pYaliG1-NLuc	pYaliG1 carrying NanoLuc luciferase (FAS1 promoter)	This study
pYaliH1-NLuc	pYaliH1 carrying NanoLuc luciferase (ICL1 promoter)	This study
pYaliJ1-NLuc	pYaliJ1 carrying NanoLuc luciferase (POX4 promoter)	This study
pYaliK1-NLuc	pYaliK1 carrying NanoLuc luciferase (G6PDH promoter)	This study
pYaliL1-NLuc	pYaliL1 carrying NanoLuc luciferase (IDP2 promoter)	This study
pYaliM1-NLuc	pYaliM1 carrying NanoLuc luciferase (GAPHD promoter)	This study
pYaliA1-B-o-N	pYaliA1 carrying EcBirA upstream of NLuc in operon configuration	This study
pYaliA1-B-p-N	pYaliA1 carrying EcBirA upstream of NLuc in pseudo-operon configuration	This study
pYaliA1-B-m-N	pYaliA1 carrying EcBirA upstream of NLuc in monocistronic configuration	This study
pYaliA1-N-o-B	pYaliA1 carrying EcBirA downstream of NLuc in operon configuration	This study
pYaliA1-N-p-B	pYaliA1 carrying EcBirA downstream of NLuc in pseudo-operon configuration	This study
pYaliA1-N-m-B	pYaliA1 carrying EcBirA downstream of NLuc in monocistronic configuration	This study
PYaliA1-hCas9	pYaliA1 carrying humanized Cas9	This study
PYaliA1-gRNA-ylCAN1	pYaliA1 carrying guide RNA targeting ylCAN1	This study
PYaliA1-hCas9-gRNA-ylCAN1	PYaliA1 carrying humanized Cas9 and guide RNA targeting ylCAN1	This study
Strains		
E. coli NEB5α	*fhuA2 Δ (argF-lacZ)U169 phoA glnV44 Φ80 Δ (lacZ)M15 gyrA96 recA1 relA1 endA1 thi-1 hsdR17*	New England Biolabs
Y. lipolytica p01g	*matA, xpr2–332, axp-2, leu2–270*	Madzak et al.

BioTek Synergy H1M microplate reader.

EmbiTech RunOne™ gel electrophoresis system.

Bio-Rad C1000™ Thermal cycler (dual block).

Bechman Allegera X-14R centrifuge.

Millipore MilliQ ultrapure water purification system (with 0.22 μm filter).

Table 2
Primers and synthetic oligos used in this protocol

No.	Primer	Nucleotide sequence (5′ > 3′)
1	VioA_F	CCGACCAGCACTTTTTGCAGTACTAACCGCAGAAGCATTC TTCCGATATCTGCATTG
2	VioA_R	GGACAGGCCATGGAACTAGTCGGTACCTCACGCGGCGA TGCGCTGCAGCAGCAG
3	VioB_F	CCGACCAGCACTTTTTGCAGTACTAACCGCAGAGCATTC TGGATTTTCCACGCATCC
4	VioB_R	GGACAGGCCATGGAACTAGTCGGTACCTCAGGCCTCTC TCGAAAGCTTTCCACAAG
5	VioC_F	CCGACCAGCACTTTTTGCAGTACTAACCGCAGAAAAGAGCAA TCATAGTCGGAGGCG
6	VioC_R	GGACAGGCCATGGAACTAGTCGGTACCTCAGTTGACCC TCCCTATCTTGTACC
7	VioD_F	CCGACCAGCACTTTTTGCAGTACTAACCGCAGAAGATTCTGG TCATCGGCGCGGG
8	VioD_R	GGACAGGCCATGGAACTAGTCGGTACCTCAGCG TTGCAGCGCGTAGCGCAGG
9	VioE_F	CCGACCAGCACTTTTTGCAGTAC TAACCGCAGGAAAACCGGGAACCGCCGCTGCTGC
10	VioE_R	GGACAGGCCATGGAACTAGTCGGTACCCTAACGC TTGGCGGCGAAGACGGCG
11	pB_F	GGCATCCCTAAATTTGATGAAAGCCTAGGGATCCATCC TCGACCAGTGGCACGCTC
12	pB_R	GCTGCCTCTGAAACTCACCATTCTAGAAGCTTTTGTTTTGTG TGACTTGTCTGTTG
13	pC_F	GGCATCCCTAAATTTGATGAAAGCCTAGGGCTCAACAAG TAGCACGTCAAATATGC
14	pC_R	GCTGCCTCTGAAACTCACCATTCTAGATCAGGAGCAGACGA TTAAATTATATCTGG
15	pD_F	GGCATCCCTAAATTTGATGAAAGCCTAGGAAATGACGACAA TTCGAACCGAGGCC
16	pD_R	GCTGCCTCTGAAACTCACCATTCTAGAGATTGATCTGGTGAG TGTCGGATTCGG
17	pE_F	GGCATCCCTAAATTTGATGAAAGCCTAGGAGTACAATTGTGC TGTGCGGATGTCC
18	pE_R	GCTGCCTCTGAAACTCACCATTCTAGATGTGATTTC TTTGGGGGTTCCACGAAG
19	pF_F	GGCATCCCTAAATTTGATGAAAGCC TAGGAGGAAACAGCGCGGCCAATTGAGGG

(continued)

Table 2
(continued)

No.	Primer	Nucleotide sequence (5′ > 3′)
20	pF_R	GCTGCCTCTGAAACTCACCATTCTAGATGTAGTTGCGTGTGG TTGTAGATTAG
21	pG_F	GGCATCCCTAAATTTGATGAAAGCCTAGGCAACTTTAG TTGGGAATATTGTTGATC
22	pG_R	GCTGCCTCTGAAACTCACCATTCTAGACGTAATGAACTG TTTGTGTGTGTAGTTC
23	pH_F	GGCATCCCTAAATTTGATGAAAGCC TAGGAAACGCGACCGGAATCTCTCGCGGTC
24	pH_R	GCTGCCTCTGAAACTCACCATTCTAGATTTTGTATGCTTGG TCAGTCTACTGGG
25	pJ_F	GGCATCCCTAAATTTGATGAAAGCCTAGGAAAGTCCGGAGA TGACACTGGGGTAAC
26	pJ_R	GCTGCCTCTGAAACTCACCATTCTAGATGTTGTGGGTCG TTTCAATGAAAGAGC
27	pK_F	GGCATCCCTAAATTTGATGAAAGCCTAGGGTTCATGCGGC TGGAGCCTCGGTTG
28	pK_R	GCTGCCTCTGAAACTCACCATTCTAGAGGTGATGTATGTA TGGTGAGTGTCTTG
29	pL_F	GGCATCCCTAAATTTGATGAAAGCCTAGGCTCCGGTTATTGG TGCATGTTAGGATG
30	pL_R	GCTGCCTCTGAAACTCACCATTCTAGAGG TTTTAGAAAGGGCAGCAAGCCGGGTGGTAGC
31	pM_F	GGCATCCCTAAATTTGATGAAAGCC TAGGGACCGGCCGGCGAGCATAAACCGGAC
32	pM_R	GCTGCCTCTGAAACTCACCATTCTAGATGTTGATGTGTG TTTAATTCAAGAATG
33	NLuc_Fw	CCGACCAGCACTTTTTGCAGTACTAACCGCAGGTCTTCACAC TCGAAGATTT
34	NLuc_Rv	CGTGGGGACAGGCCATGGAACTAGTCGGTACC TTACGCCAGAATGCGTTCGC
35	TEF_F	GCAGTCTGGAATCTACGCTTGTTCA
36	XPR2_R	GCCTAAATGCCAACACCGTTGTAGGC
37	SeqPro_F	CTGTGAGGCGTTAAATAGAGCTTG
38	hCas9_F	catccgaccagcactttttgcagtactaaccgcaggacaagaagtactccattgggctcg
39	hCas9_R	GGACAGGCCATGGAACTAGTCGGTACCTCACACCTTCCTC TTCTTCTTGGGGTCAG
40	ylCan1_gRF	ctcgtcaccatcaacttcttccccgtgttttagagctagaaatagcaag

(continued)

Table 2
(continued)

No.	Primer	Nucleotide sequence (5′ > 3′)
41	ylCan1_gRR	ACGGGGAAGAAGTTGATGGTGACGAGCTTACTC
42	Can1_seq	ccggtctcttcattggtacc
43	Can1_F	gaaaagacattttcaaacgattacccacc
44	Can1_R	GGATGAGAGTGATGAAGATGATGG
45	HHR-gRNA-Can1-tracrRNA-HDV	GCATTTCCTTCTGAGTATAAGAATCATTCAAATCTAGAATGGTGAGTTTCAGAGGCAGCAGCAATTGCCACGGGCTTTGAGCACACGGCCGGGTGTGGTCCCATTCCCATCGACACAAGACGCCACGTCATCCGACCAGCACTTTTTGCAGTACTAACCGCAGGATGGTCTGATGAGTCCGTGAGGACGAAACGAGTAAGCTCGTCACCATCAACTTCTTCCCCGTGTTTTAGAGCTAGAAATAGCAAGTTAAAATAAGGCTAGTCCGTTATCAACTTGAAAAAGTGGCACCGAGTCGGTGCTTTTGGCCGGCATGGTCCC

3 Methods

3.1 Y. lipolytica Transformation

1. Prior to starting the transformation, prepare a YPD plate streaked with the appropriate host strain of *Y. lipolytica* and incubate overnight.
2. Prepare a master mix of yeast transformation buffer solution. ssDNA should be boiled at 95 °C for at least 3 min. The table below shows the components for one reaction.

Component	Volume (μL)
Sterile DI water	10
Sterile 50% PEG4000	80
2 M lithium acetate	5
Boiled ssDNA (salmon sperm, denatured)	5
Total volume	100

3. Scrape a lawn of *Y. lipolytica* P01gΔleu strain from the YPD plate and transfer to each reaction of transformation buffer. Mix well by vortexing for 10 s.
4. Add 0.25–0.5 μg of plasmid DNA, mix thoroughly by vortexing for 10 s.
5. Incubate the transformation mixture at 30 °C for 30–45 min, with vortexing the mixture occasionally every 10 min for 10 s. Prepare a water bath at 39 °C.

6. Heat shock cells at 39 °C for 10 min.
7. Dilute the cell culture 4× by adding 300 μL of sterile water. Plate 200 μL of the diluted culture on yeast CSM selective plates with the appropriate dropout component. Incubate the plate at 30 °C for at least 2 days. Incubate remaining cell culture overnight at 30 °C and 250 rpm to enrich for successful transform ants.

3.2 Preparation of Plasmids, Genomic DNA and PCR Fragments

1. Inoculate *E. coli* NEB5α harboring the plasmid pYaliA1-NLuc in 3 mL LB medium supplemented with 100 μg/mL of ampicillin in a snap-cap 12 mL culture tube. Incubate overnight at 37 °C, 250 rpm.
2. Repeat **step 1** with the *E. coli* NEB5α strain harboring pYaliA1-EcBirA.
3. Inoculate *Y. lipolytica* strain P01g in 3 mL YPD medium in a 12 mL culture tube. Incubate overnight at 30 °C, 250 rpm.
4. Centrifuge overnight cultures at 4,400 × *g* for 10 min. Discard supernatant.
5. Using the Zyppy™ Plasmid Miniprep kit, extract plasmid DNA from *E. coli* strains and elute plasmids from the column with 25 μL RNase- and DNase-free water. Repeat this elution step once more. Store plasmid preparations at −20 °C for future use.
6. To extract genomic DNA from *Y. lipolytica* strain P01g, use the Promega Wizard Genomic DNA purification Kit. Store genomic DNA at −20 °C for future use.
7. Set up PCR reactions to amplify the following promoters (Table 3) from *Y. lipolytica* genomic DNA:

Component	Volume (μL)
10 μM Forward primer	1.25
10 μM Reverse primer	1.25
DNA template (gDNA)	2.5
Q5 High-Fidelity 2× Master Mix	12.5
DNase/RNase-free water	7.5
Total volume	25

Table 3
Promoters cloned in this study

Transcriptional unit annotation	Transcriptional unit	Promoter symbol	Vector name
Translational elongation factor EF-1 alpha	YALI0C09141g	ylTEF promoter	pYaliA1
Acyl-CoA:diacylglycerol acyltransferase	YALI0E32769g	ylDGA1 promoter	pYaliB1
Acetyl-CoA-carboxylase 1	YALI0C11407g	ylACC promoter	pYaliC1
ATP citrate lyase 2	YALI0D24431g	ylACL2 promoter	pYaliD1
Isocitrate dehydrogenase NAD+ subunit 2 mitochondrial	YALI0D06303g	ylIDH2 promoter	pYaliE1
Fatty acid synthase subunit beta	YALI0B15059g	ylFAS2 promoter	pYaliF1
Fatty acid synthase subunit alpha	YALI0B19382g	ylFAS1 promoter	pYaliG1
Isocitrate lyase 1	YALI0C16885g	ylICL1 promoter	pYaliH1
Fatty-acyl coenzyme A oxidase	YALI0E27654g	ylPOX4 promoter	pYlaliJ1
Glucose-6-phosphate dehydrogenase	YALI0E22649g	ylG6PDH promoter	pYaliK1
Cytosolic NADP-specific isocitrate dehydrogenase	YALI0F04095g	ylIDP2 promoter	pYaliL1
Glyceraldehyde 3-phosphate dehydrogenase	YALI0C06369g	ylGAPHD promoter	pYaliM1

8. Perform the PCR reactions using the following parameters:

Steps	Cycles	Temperature	Time (mm:ss)
Initial denaturation	1	98 °C	00:30 (plasmid) or 01:30 (genomic DNA)
Denaturation	25–35	98 °C	00:10 (plasmid) or 00:30 (genomic DNA)
Annealing		50–68 °C	00:25
Extension		72 °C	00:30 per kb
Additional extension	1	72 °C	05:00
Storage	1	16 °C	As needed

9. Run a DNA electrophoresis gel to check for correct fragment sizes and perform DNA purification using the Zymo DNA Clean & Concentrator™-25 kit.

10. Repeat **steps 7–9** but with genomic DNA obtained from *C. violaceum* to amplify the vioA, vioB, vioC, vioD, vioE genes and obtain purified PCR fragments.

3.3 Assembly of 12 Promoters in Yalibrick Vectors

1. Perform double digestion on the plasmid pYaliA1-NLuc using the restriction enzymes AvrII and XbaI. Digest approximately 0.5 μg of plasmid in a 30 μL digestion reaction, scaling it up accordingly.
2. Run an electrophoresis gel to extract the destination vector fragment. The expected fragment size is 7534 bp.
3. Set up Gibson assembly reactions to clone the other 11 promoters (Table 3) into the pYaliBrick vector. Incubate the reactions for 1 h at 50 °C in a thermocycler block.

Component	Volume (μL)
Insert	3 (150–300 ng)
Vector	2 (50–150 ng)
2× Gibson mix	7.5
DNase/RNase-free water	2.5
Total volume	15

4. Immediately transfer the tubes onto ice. Thaw 11 shots of NEBα High Efficiency chemically competent cells. Perform transformation as outlined in Subheading 3.1.
5. After recovering the cells, plate 80 μL of each cell culture on a LB plates with 100 μg/mL ampicillin. Incubate at 37 °C for 18 h.
6. Pick at least three colonies from each plate to perform colony PCR to screen for positives. The primer pair SeqPro_F and XPR2_R are used for all constructs.
7. Run a gel to visualize to results and identify positives for inoculation and miniprep.
8. After obtaining the plasmid DNA from overnight cultures, sequence the constructs for verification.

3.4 Characterization of Promoter Library with NanoLuc® Luciferase

1. Transform the 12 promoter constructs into *Y. lipolytica* P01g ΔLeu as outlined in Subheading 3.2 and plate on CSM-Leu selective plates.
2. Select triplicate colonies from the grown plates and inoculate in 3 mL CSM-Leu in culture tubes for shake flask seed. Incubate at least 36 h at 30 °C and 250 rpm.
3. Start shake flask cultures using the seed. For 50 mL of working volume in a 250 mL shake flask, inoculate with initial OD_{600} ~0.01. Incubate shake flasks at 30 °C and 250 rpm for 18 h.
4. Harvest enough cell culture from each sample to obtain 50 mL of $OD_{600} = 1$ amount of cells. Transfer to 50 mL Falcon tubes and centrifuge at 4,400 × *g*, 10 min. Discard supernatant and freeze cell pellets at −80 °C for storage.

5. To prepare cell lysates, thaw frozen cell pellets on ice. Prepare cell lysis buffer as follows:

Component	Volume (mL)
Luciferase Cell Culture Lysis 5× Reagent	2
26 mg/mL Zymolyase	0.38
Sterile DI water	8
1 Roche mini EDTA-free protease inhibitor cocktail tablet	–
Total volume	10

6. Resuspend each cell pellet in 500 μL of lysis buffer and incubate at room temperature for 5 min.
7. Transfer each reaction into a 1.5 mL centrifuge tube containing 250 μL of 0.5 mm glass beads. Vortex for 2 min to shear the cells.
8. Centrifuge the samples in a temperate-controlled microcentrifuge at 4 °C for 30 s, 14,000 × *g*. Transfer the supernatant to a new centrifuge tube on ice.
9. Add equal volume of cold 75% glycerol to each sample and mix well. Measure total protein concentration with the Bradford assay kit.
10. To perform the luciferase assay, equilibrate the Nano-Glo® Luciferase Assay Buffer to room temperature. Normalize each lysate sample to 0.04 mg/mL total protein concentration to prevent oversaturation of the microtiter plate detector. Use 1× PBS buffer for dilution.
11. Each reaction in a 96-well white microtiter plate requires the following:

Component	Volume (μL)
Nano-Glo® Luciferase Assay Buffer	100
Nano-Glo® Luciferase Assay Substrate	2
Lysate (diluted to 0.04 mg/mL)	100
Total volume	202

12. It is recommended to make a master mix of the assay buffer and substrate first. Aliquot the lysates into the wells.
13. Aliquot 100 μL of the luciferase assay master mix with a multichannel pipettor and immediately transfer the microtiter place to the plate reader to measure RLU over a time course of 15 min at 30 s intervals.

3.5 Assembly of Two-Gene Pathway in Different Genetic Configurations

1. EcBirA and NLuc can be assembled in the pYaliA1 backbone to give a total of 6 different configurations as seen in Fig. 1.
2. Using the plasmids pYaliA1-EcBirA and pYaliA1-NLuc, appropriate restriction enzyme sites are selected in order to determine gene positional order and configuration (Fig. 2). The tables below illustrate the composition for each reaction of restriction digestion and the expected fragment sizes. Desired fragment sizes are indicated in underlined bold. Set up the digestion reactions accordingly and incubate for 30–45 min at 37 °C. The first table shows the configuration that EcBrirA is assembled in front of Nluc. The second table shows the configuration that Nluc is assembled in front of EcBirA.

Insert (Donor vector pYaliA1-Nluc)				Vector (Destination vector pYaliA1-EcBirA)			
pYaliA1-Nluc	5 μL	pYaliA1-Nluc	5 μL	pYaliA1-EcBirA	3 μL	pYaliA1-EcBirA	3 μL
XbaI	2 μL	AvrII	2 μL	SpeI	2 μL	NheI	2 μL
SalI	2 μL	SalI	2 μL	SalI	2 μL	SalI	2 μL
10 × FastDigest Buffer	3 μL	10× FastDigest Buffer	3 μL	10× FastDigest Buffer	3 μL	10× FastDigest Buffer	3 μL
Water	18 μL	Water	18 μL	Water	20 μL	Water	20 μL
Expected fragments (bp):	6836 + **1116**	Expected fragments (bp):	6418 + **1534**	Expected fragments (bp):	**7937** + 465	Expected fragments (bp):	**8379** + 23

Insert (Donor vector pYaliA1-EcBirA)				Vector (Destination vector pYaliA1-Nluc)			
pYaliA1-EcBirA	5 μL	pYaliA1-EcBirA	5 μL	pYaliA1-NLuc	3 μL	pYaliA1-NLuc	3 μL
XbaI	2 μL	AvrII	2 μL	SpeI	2 μL	NheI	2 μL
SalI	2 μL	SalI	2 μL	SalI	2 μL	SalI	2 μL
10 × FastDigest Buffer	3 μL	10× FastDigest Buffer	3 μL	10× FastDigest Buffer	3 μL	10× FastDigest Buffer	3 μL
Water	18 μL	Water	18 μL	Water	20 μL	Water	20 μL
Expected fragments (bp):	6836 + **1566**	Expected fragments (bp):	6418 + **1984**	Expected fragments (bp):	**7487** + 465	Expected fragments (bp):	**7929** + 23

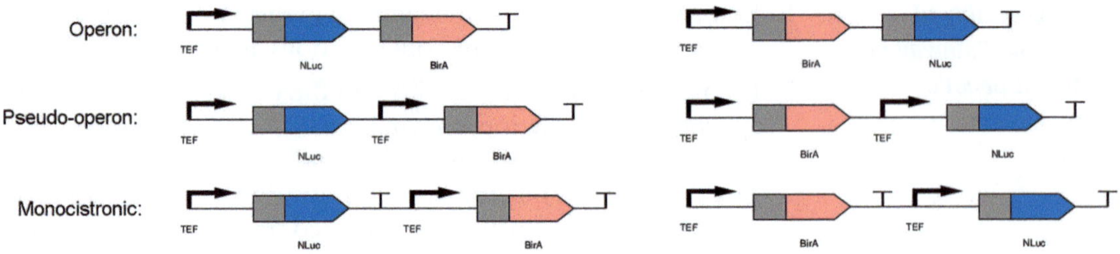

Fig. 1 All possible combinations of a two-gene pathway in a pYalibrick vector. The grey blocks indicate intronic sequence fused upstream of each gene. The operon and pseudo-operon configuration relies on the alternative-splicing of mRNA for functional expression of two tandem gene

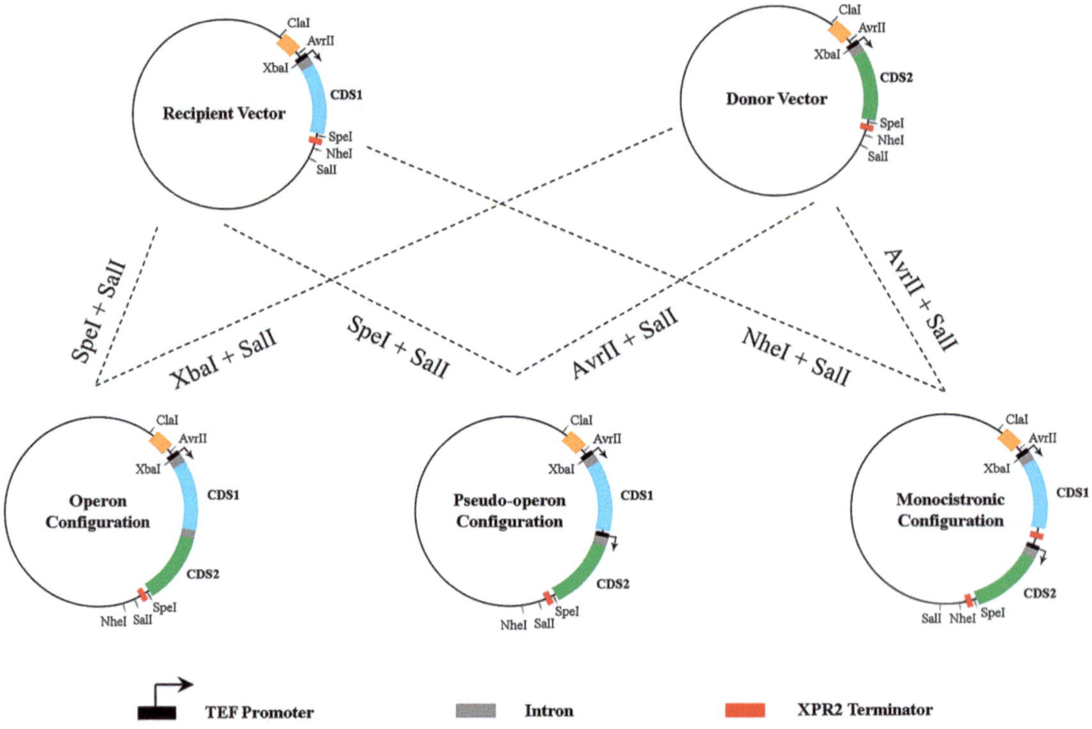

Fig. 2 Yalibrick cloning strategy for constructing the three different gene configurations: operon, pseudo-operon, monocistronic. The use of isocaudomers allow for repeated use of the same restriction sites for subsequent cloning

3. Run an electrophoresis gel to check the digestion results and extract the desired fragments with Zymo clean gel purification kits.
4. Set up ligation reactions with T4 ligase. Ideally, each 10 μL reaction should contain at least 25 ng of vector and 3–5 times the molar quantity of insert. Incubate the ligation reaction at room temperature for at least 1 h. The required fragments for each construct can be obtained by the following combinations:

Plasmid	Fragments
pYaliA1-B-o-N	NLuc (XbaI-SalI) + EcBirA (SpeI-SalI)
pYaliA1-B-p-N	NLuc (AvrII-SalI) + EcBirA (SpeI-SalI)
pYaliA1-B-m-N	NLuc (AvrII-SalI) + EcBirA (NheI-SalI)
pYaliA1-N-o-B	EcBirA (XbaI-SalI) + NLuc (SpeI-SalI)
pYaliA1-N-p-B	EcBirA (AvrII-SalI) + NLuc (SpeI-SalI)
pYaliA1-N-m-B	EcBirA (AvrII-SalI) + NLuc (NheI-SalI)

5. Use 4 μL of ligation reaction to transform into 10 μL NEBα High Efficiency chemically competent cells as outlined in Subheading 3.1. Plate 80 μL of recovered cell culture on 100 μg/mL ampicillin LB plates. Incubate at 37 °C for 18 h.
6. Select two colonies for each construct. Inoculate each colony in 3 mL LB + 100 μg/mL ampicillin. Incubate at 37 °C overnight.
7. Using the Zyppy™ Plasmid Miniprep kit, extract plasmid DNA from the overnight cultures and elute plasmids from the column with 25 μL RNase and DNase-free water. Repeat this elution step once more.
8. Screen the plasmids by performing double restriction digestion with the restriction enzymes KpnI and XhoI. The reaction mixture can be scaled down to 15 μL.
9. Run a DNA electrophoresis gel to screen the plasmids. A detailed digestion pattern of the six gene configuration is illustrated in Fig. 3.

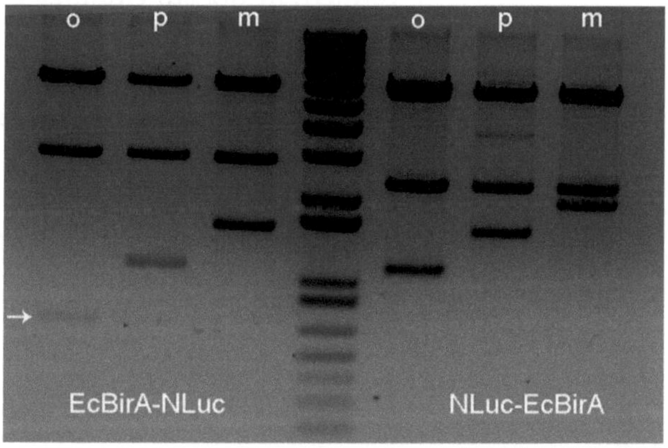

Fig. 3 Digestion screening results of pYaliA1-EcBirA-NLuc and pYaliA1-NLuc-EcBirA with KpnI and XhoI restriction enzymes. A distinct stepwise pattern is seen in the smallest fragment from the resulting digestion reaction, indicating the addition of functional promoter and terminator

3.6 Assembly and Verification of Five-Gene Violacein Biosynthetic Pathway

1. Inoculate *E. coli* NEB5α harboring the plasmid pYaliA1 in 3 mL LB medium supplemented with 100 μg/mL of ampicillin in a snap-cap 12 mL culture tube. Incubate overnight at 37 °C, 250 rpm.
2. Extract the plasmid DNA from the overnight culture via mini-prep. Elute with 50 μL of RNase and DNase-free water.
3. Perform double restriction enzyme digestion on pYaliA1 plasmid with KpnI and SnaBI to obtain the destination vector. Run a gel to extract the fragment.
4. Set up Gibson assembly reactions using the PCR fragments of vioA, vioB, vioC, vioD, and vioE obtained from Subheading 3.3. To each reaction mix, add linearized pYaliA1 (KpnI-SnaB1) as the destination vector and one of the five PCR fragments to obtain the following five constructs: pYaliA1-vioA, pYaliA1-vioB, pYaliA1-vioC, pYaliA1-vioD, and pYaliA1-vioE.
5. Transform 4 μL of each Gibson assembly mixture into NEB5α High Efficiency chemically competent cells as outlined in Subheading 3.1.
6. After recovering the cell cultures, plate 80 μL for each reaction of transformant on LB plates with 100 μg/mL ampicillin. Incubate the plates at 37 °C for 18 h.
7. Screen colonies from the 5 plates using colony PCR and the primer pair TEF_F and XPR2_R. Run a gel to visualize the results and select 2 positives from each construct for further verification via sequencing.
8. Upon sequence verification of all five constructs, proceed with the first round of subcloning. Since the sequences has been verified, future subclonings to not require additional sequence verification.
9. The first round of subcloning involves cloning a second gene into the vector that already contains one gene. The expected constructs to be obtained in this first round are pYaliA1-vioB-vioA and pYaliA1-vioD-vioC. A detailed procedure to assemble the five-gene violacein pathway is illustrated in Fig. 4.
10. Set up the following digestion reactions:

Construct: pYaliA1-vioB-vioA				**Construct: pYaliA1-vioD-vioC**			
pYaliA1-vioA	3	pYaliA1-vioB	3	pYaliA1-vioC	3	pYaliA1-vioD	3
AvrII	2	NheI	2	AvrII	2	NheI	2
ClaI	2	ClaI	2	SalI	2	SalI	2
10× FastDigest Buffer	3	10× FastDigest Buffer	3	10× FastDigest Buffer	3	10× FastDigest Buffer	3
Water	20	Water	20	Water	20	Water	20
Expected fragments:	232 + **8461**	Expected fragments:	**4224** + 6209	Expected fragments:	**2308** + 6418	Expected fragments:	**8535** + 23

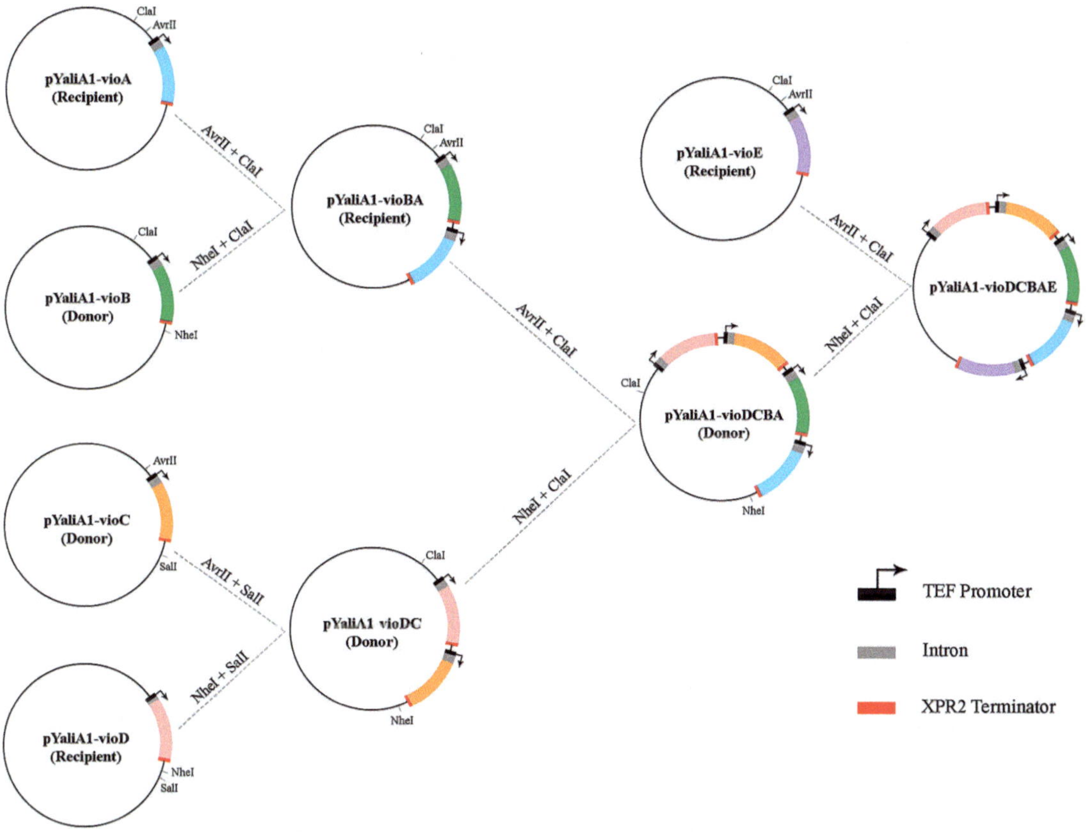

Fig. 4 Sequential cloning schematic for the assembly of the five-gene pathway of violacein pathway in pYaliA1 backbone

11. Run an electrophoresis gel to purify the desired fragments (highlighted in bold). Ligate the fragment pairs accordingly using the protocol indicated in **step 4** of Subheading 3.6.
12. Transform the ligated products into NEB5α High Efficiency chemically competent cells as outlined in Subheading 3.1.
13. After recovering the cell cultures, plate 80 μL for each reaction of transformant on LB plates with 100 μg/mL ampicillin. Incubate the plates at 37 °C for 18 h.
14. Pick at least three colonies for each construct and inoculate in 3 mL of LB with 100 μg/mL ampicillin. Incubate overnight at 37 °C.
15. Miniprep the overnight cultures to extract the plasmid DNA. Perform restriction digestion on the plasmids to screen for positive colonies by cutting with KpnI and XhoI. The expected fragment sizes for pYaliA1-vioB-vioA and pYaliA1-vioD-vioC are 2252, 5654, 29, 4750 bp and 2285, 5654, 29, 2875 bp respectively.
16. Positive clones of pYaliA1-vioB-vioA and pYaliA1-vioD-vioC are subsequently used for the second subcloning to generate the construct pYaliA1-vioDCBA.
17. Set up the digestion reaction as follows:

Construct: pYaliA1-vioDCBA			
pYaliA1-vioBA	3	pYaliA1-vioDC	3
ClaI	2	ClaI	2
AvrII	2	NheI	2
10× FastDigest Buffer	3	10× FastDigest Buffer	3
Water	20	Water	20
Expected fragments:	**12,453** + 232	Expected fragments:	6209 + **4634**

18. Run an electrophoresis gel to purify the desired fragments (highlighted in bold). Ligate the fragment pair accordingly using the protocol indicated in **step 4** of Subheading 3.6.
19. Transform the ligated product into NEB5α High Efficiency chemically competent cells as outlined in Subheading 3.1.
20. After recovering the cell culture, plate 80 μL of transformant on LB plates with 100 μg/mL ampicillin. Incubate the plate at 37 °C for 18 h.
21. Pick at least five colonies and inoculate in 3 mL of LB with 100 μg/mL ampicillin. Incubate overnight at 37 °C.
22. Miniprep the overnight cultures to extract the plasmid DNA. Perform restriction digestion on the plasmids to screen for

positive colonies by cutting with HindIII and EcoRI. The expected fragment sizes are 3866, 849, 6608, 1335, 962, 3467 bp.

23. Positive clone of pYaliA1-vioDCBA is subsequently used for the final subcloning to generate the construct pYaliA1-vioDCBAE.

24. Set up the digestion reaction as follows:

Construct: pYaliA1-vioDCBAE			
pYaliA1-vioE	3	pYaliA1-vioDCBA	3
AvrII	2	NheI	2
ClaI	2	ClaI	2
10× FastDigest Buffer	3	10× FastDigest Buffer	3
Water	20	Water	20
Expected fragments:	232 + **7780**	Expected fragments:	**10,878** + 6209

25. Run an electrophoresis gel to purify the desired fragments (highlighted in bold). Ligate the fragment pairs accordingly using the protocol indicated in **step 4** of Subheading 3.6.

26. Transform the ligated product into NEB5α High Efficiency chemically competent cells as outlined in Subheading 3.1.

27. After recovering the cell culture, plate 80 μL of transformant on LB plates with 100 μg/mL ampicillin. Incubate the plate at 37 °C for 18 h.

28. Pick at least five colonies for each construct and inoculate in 3 mL of LB with 100 μg/mL ampicillin. Incubate overnight at 37 °C.

29. Miniprep the overnight cultures to extract the plasmid DNA. Perform restriction digestion on the plasmids to screen for positive colonies by cutting with EcoRI. The expected fragment sizes are 3866, 9028, 1335, 962, 3467 bp.

30. Transform the five gene vector pYaliA1-VioDCBAE into *Y. lipolytica* P01g Δleu, follow the protocol as described in Subheading 3.1.

31. Grow the plates at 30 °C for 2 or 3 days. A dark blue lawn of yeast colony was formed demonstrating the functional expression of the five-gene violacein pathway (Fig. 5).

Fig. 5 Functional expression of five-gene violacein pathway in oleaginous yeast *Yarrowia lipolytica*

3.7 Genome Editing with CRISPR-Cas9 and Hammer-Head Ribozyme Delivered CAN1 gRNA

1. Digest plasmid pYaliA1 with SnaBI and KpnI, incubate at 37 °C water bath for 1 h. Gel purification of digested vector pYaliA1 with Zymo clean gel extraction kit.
2. Digest plasmid pYaliA1 with XbaI and SpeI, incubate at 37 °C water bath for 1 h. Gel purification of digested vector pYaliA1 with Zymo Clean Gel extraction kit.
3. PCR amplify hCas9 from p414-TEF-hCas9 (Addgene) with primer hCas9_F and hCas9_R. Digest the PCR products with DpnI to remove template DNA, then the PCR product is cleaned up and concentrated with Zymo DNA Clean & Concentrator Kits.
4. Assemble SnaB1 and KpnI digested pYaliA1 vector backbone with cleaned PCR products hCas9 with Gibson Assembly kits. Transform 2.5 μL Gibson reaction to NEB5α high efficiency cell. Heat-shock and screen/sequence positive trans formats with standard protocol. The positive clone is named pYaliA1-hCas9.
5. Assemble XbaI and SpeI digested pYaliA1 vector backbone with synthetic gene HHR-gRNA-Can1-tracrRNA-HDV with Gibson Assembly Kits. Transform 2.5 μL Gibson reaction to NEB5α high efficiency cell. Heat shock and screen/sequence positive transformats with standard protocol. The positive clone is named pYaliA1-gRNA-ylCAN1.
6. Digest donor vector pYaliA1-gRNA-ylCAN1with AvrII and NotI, digest destination vector pYaliA1-hCas9 with NheI and Not1. Gel purified gene fragments were ligated with NEB T4 DNA ligase and 2.5 μL ligation reaction was transformed to NEB5α high efficiency cell. Heat shock and screen/sequence

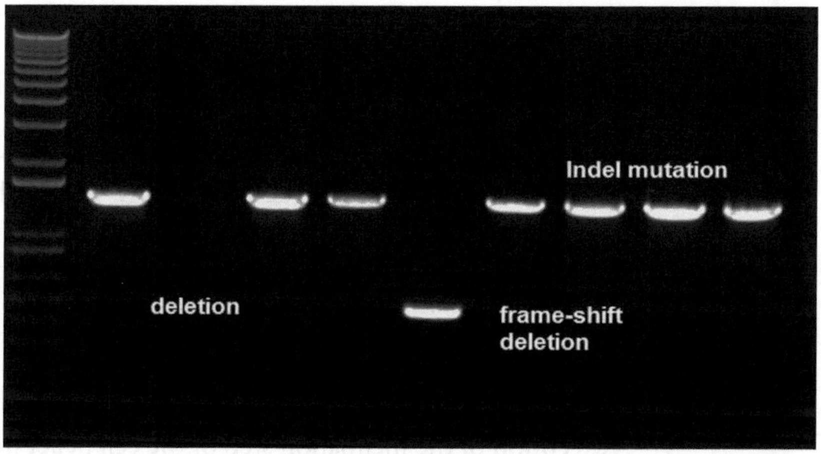

Fig. 6 Gene deletion, frameshift deletion, and indel mutation is screened by gel electrophoresis of PCR fragment of CAN1 gene. The indel mutation and frameshift deletion was later verified by Sanger sequencing

positive transformats with standard protocol. The positive clone is named pYaliA1-hCas9-gRNA-ylCAN1.

7. Transform 0.5 μg of pYaliA1-hCas9-gRNA-ylCAN1 into *Y. lipolytica* P01g Δleu, follow the protocol as described in Subheading 3.1.
8. Plate half of the transformation solution onto CSM-Arg (yeast complete synthetic media without Arginine) supplied with 50 μg/mL canavanine. Grow at 30 °C incubator for 2 or 3 days.
9. Pick isolated colonies and grow in 2 mL YPD media at 30 °C with shaking at 250 rpm overnight.
10. Purify *Yarrowia lipolytica* genomic DNA with the Promega Wizard genomic DNA purification kits.
11. From each genomic DNA sample, amplify CAN1 gene with the primer pairs Can1_F and Can1_R. The PCR products was analyzed by Gel and later was cleaned-up and sent to Genewiz for DNA sequencing with Can1_seq as sequencing primer. An example of indel mutation and frameshift gene deletion of CAN1 is illustrated in Fig. 6.

4 Notes

1. *E. coli* transformation: If no colonies are obtained from the first 80 μL that was plated right after recovering the cell culture, plate the remaining cell culture that was left incubating at room temperature overnight.
2. *Y. lipolytica* transformation: If no colonies are seen from the first round of plating, plate the remaining overnight culture on a fresh selective CSM plate.

3. Save the ligation reaction mixture after using it for transformation by keeping it at room temperature overnight. If there are no colonies after plating all the culture, redo the transformation with this overnight ligation mixture.
4. Sequence all recombinant DNA construct if the insert is obtained from PCR. For gene assembly via subcloning, it is not necessary to sequence after every step.
5. The volume of glass beads used in the mechanical lysing of Yarrowia should not exceed more than one-third of the total volume. Lysing may not be more effective at higher volumes of glass beads.
6. The entire process of cell lysis is performed at 4 °C with the exception of the incubation step of the cell pellet in lysis buffer at room temperature.
7. For YaliBrick gene assembly, the insertion fragment should be free of AvrII, XbaI, SpeI, NheI and SalI (or ClaI) restriction sites. If any of these sites appears in the insertion fragment, site-directed mutagenesis must be used to silently mutate these sites.
8. The CRISPR-Cas9 demonstrated relatively low on-target efficiency, about 80% of the mutant were off-target indel mutation of CAN1 gene, 10% of the mutant were frameshift deletion, and 10% of the mutant were on-target mutation of the designed PAM sequence.

References

1. Xu P et al (2016) Engineering Yarrowia lipolytica as a platform for synthesis of drop-in transportation fuels and oleochemicals. Proc Natl Acad Sci U S A 113(39):10848–10853
2. Abghari A, Chen S (2014) Yarrowia lipolytica as an oleaginous cell factory platform for the production of fatty acid-based biofuel and bioproducts. Front Ener Res 2(21):1–21
3. Ledesma-Amaro R, Nicaud J-M (2016) Yarrowia lipolytica as a biotechnological chassis to produce usual and unusual fatty acids. Prog Lipid Res 61:40–50
4. Xu P, Qiao K, Stephanopoulos G (2017) Engineering oxidative stress defense pathways to build a robust lipid production platform in Yarrowia lipolytica. Biotechnol Bioeng 114 (7):1521–1530
5. Zhu Q, Jackson EN (2015) Metabolic engineering of Yarrowia lipolytica for industrial applications. Curr Opin Biotechnol 36:65–72
6. Ledesma-Amaro R et al (2016) Metabolic engineering of Yarrowia lipolytica to produce chemicals and fuels from xylose. Metab Eng 38:115–124
7. Beopoulos A, Chardot T, Nicaud JM (2009) Yarrowia lipolytica: a model and a tool to understand the mechanisms implicated in lipid accumulation. Biochimie 91(6):692–696
8. Fickers P et al (2005) Hydrophobic substrate utilisation by the yeast Yarrowia lipolytica, and its potential applications. FEMS Yeast Res 5 (6–7):527–543
9. Ruiz-Herrera J, Sentandreu R (2002) Different effectors of dimorphism in Yarrowia lipolytica. Arch Microbiol 178:477–483
10. Morales-Vargas AT, Domínguez A, Ruiz-Herrera J (2012) Identification of dimorphism-involved genes of Yarrowia lipolytica by means of microarray analysis. Res Microbiol 163(5):378–387
11. Beckerich JM, Boisramé-Baudevin A, Gaillardin C (1998) Yarrowia lipolytica: a model organism for protein secretion studies. Int Microbiol 1:123–130

12. Groenewald M et al (2014) Yarrowia lipolytica: safety assessment of an oleaginous yeast with a great industrial potential. Crit Rev Microbiol 40(3):187–206
13. Papanikolaou S et al (2002) Yarrowia lipolytica as a potential producer of citric acid from raw glycerol. J Appl Microbiol 92(4):737–744
14. Papanikolaou S et al (2008) Citric acid production by Yarrowia lipolytica cultivated on olive-mill wastewater-based media. Bioresour Technol 99(7):2419–2428
15. Zhou J et al (2010) Screening of a thiamine-auxotrophic yeast for α-ketoglutaric acid overproduction. Lett Appl Microbiol 51 (3):264–271
16. Cui Z et al (2017) Engineering of unconventional yeast Yarrowia lipolytica for efficient succinic acid production from glycerol at low pH. Metab Eng 42:126–133
17. Xue Z et al (2013) Production of omega-3 eicosapentaenoic acid by metabolic engineering of Yarrowia lipolytica. Nat Biotechnol 31:734–740
18. Xie D, Jackson EN, Zhu Q (2015) Sustainable source of omega-3 eicosapentaenoic acid from metabolically engineered Yarrowia lipolytica: from fundamental research to commercial production. Appl Microbiol Biotechnol 99 (4):1599–1610
19. Zhang B et al (2013) Genetic engineering of Yarrowia lipolytica for enhanced production of trans-10, cis-12 conjugated linoleic acid. Microb Cell Factories 12(1):70
20. Qiao K et al (2017) Lipid production in Yarrowia lipolytica is maximized by engineering cytosolic redox metabolism. Nat Biotechnol 35(2):173–177
21. Blazeck J et al (2014) Harnessing Yarrowia lipolytica lipogenesis to create a platform for lipid and biofuel production. Nat Commun 5:3131
22. Liu L, Alper HS (2014) Draft genome sequence of the oleaginous yeast Yarrowia lipolytica PO1f, a commonly used metabolic engineering host. Genome Announc 2(4): e00652–e00614
23. Cao X et al (2016) Metabolic engineering of oleaginous yeast Yarrowia lipolytica for limonene overproduction. Biotechnol Biofuels 9 (1):214
24. Braga A, Belo I (2016) Biotechnological production of γ-decalactone, a peach like aroma, by Yarrowia lipolytica. World J Microbiol Biotechnol 32(10):169
25. Gomes N, Teixeira JA, Belo I (2012) Fed-batch versus batch cultures of Yarrowia lipolytica for γ-decalactone production from methyl ricinoleate. Biotechnol Lett 34 (4):649–654
26. Hong S-P et al (2012) Engineering Yarrowia lipolytica to express secretory invertase with strong FBA1IN promoter. Yeast 29(2):59–72
27. Madzak C (2015) Yarrowia lipolytica: recent achievements in heterologous protein expression and pathway engineering. Appl Microbiol Biotechnol 99(11):4559–4577
28. Kretzschmar A et al (2013) Increased homologous integration frequency in Yarrowia lipolytica strains defective in non-homologous end-joining. Curr Genet 59(1):63–72
29. Fickers P et al (2003) New disruption cassettes for rapid gene disruption and marker rescue in the yeast Yarrowia lipolytica. J Microbiol Methods 55(3):727–737
30. Schwartz CM et al (2016) Synthetic RNA polymerase III promoters facilitate high-efficiency CRISPR-Cas9-mediated genome editing in Yarrowialipolytica. ACS Synth Biol 5:356–359
31. Gao S et al (2016) Multiplex gene editing of the Yarrowia lipolytica genome using the CRISPR-Cas9 system. J Ind Microbiol Biotechnol 43(8):1085–1093
32. Liu L et al (2014) Increasing expression level and copy number of a Yarrowia lipolytica plasmid through regulated centromere function. FEMS Yeast Res 14:1124–1127
33. Xu P, Bhan N, Koffas MAG (2013) Engineering plant metabolism into microbes: from systems biology to synthetic biology. Curr Opin Biotechnol 24(2):291–299
34. Xu P, Koffas MG (2013) Assembly of multi-gene pathways and combinatorial pathway libraries through ePathBrick vectors. In: Polizzi KM, Kontoravdi C (eds) Synthetic biology. Humana Press, New York, pp 107–129
35. Xu P et al (2012) ePathBrick: a synthetic biology platform for engineering metabolic pathways in E-coli. ACS Synth Biol 1(7):256–266
36. Xu P et al (2017) Improving metabolic pathway efficiency by statistical model-based multivariate regulatory metabolic engineering. ACS Synth Biol 6(1):148–158
37. Xu P et al (2011) Genome-scale metabolic network modeling results in minimal interventions that cooperatively force carbon flux towards malonyl-CoA. Metab Eng 13(5):578–587
38. Xu P et al (2013) Modular optimization of multi-gene pathways for fatty acids production in *E. coli*. Nat Commun 4:1409
39. Xu P et al (2014) Improving fatty acids production by engineering dynamic pathway regulation and metabolic control. Proc Natl Acad Sci U S A 111(31):11299–11304
40. Xu P et al (2014) Design and kinetic analysis of a hybrid promoter-regulator system for Malonyl-CoA sensing in *Escherichia coli*. ACS Chem Biol 9(2):451–458

Chapter 12

Gas Chromatography–Mass Spectrometry Microbial Metabolomics for Applications in Strain Optimization

McKenzie L. Smith, April M. Miguez, and Mark P. Styczynski

Abstract

Metabolomics is the systems-scale measurement of biochemical intermediates in biological systems; by virtue of its deep and broad study of metabolism, it has great potential for applications in metabolic engineering. While a number of the analytical techniques used widely in metabolomics are familiar to metabolic engineers performing post hoc analyses of product titers, the requirements for accurately capturing metabolism at a systems scale rather than just measuring a single secreted product are much more complicated. Nonetheless, metabolomics (which is still not widely available as an affordable consumer service like many molecular biology services) is within reach of many properly equipped metabolic engineering groups. To this end, we present a detailed metabolomics protocol with application to strain optimization. Specifically, we focus on characterizing metabolism in the yeast *Saccharomyces cerevisiae* using gas chromatography coupled to mass spectrometry. The measurement of metabolic intermediates that results from such approaches has the potential to enable more informed and rational efforts towards pathway engineering and strain optimization.

Key words Metabolomics, Metabolic engineering, Strain optimization, Gas chromatography, Mass spectrometry, Metabolites, Yeast, *Saccharomyces cerevisiae*, Quenching, Extraction

1 Introduction

Metabolomics is a systems-scale approach to studying metabolism that has seen increasingly wide application in recent years and that offers great potential in a metabolic engineering context. Characterization of systems-scale metabolic trends and dynamics can help to identify key pathways to target for creation of an optimal strain phenotype for a given target molecule. Work in the field of metabolomics can make use of a wide variety of techniques, depending on available equipment and the needs of the specific application. Generally speaking, a metabolomics analysis pipeline includes a sample acquisition and preparation stage, a data acquisition stage, and a data processing stage; here, we will present an example of

Christine Nicole S. Santos and Parayil Kumaran Ajikumar (eds.), *Microbial Metabolic Engineering: Methods and Protocols*, Methods in Molecular Biology, vol. 1927, https://doi.org/10.1007/978-1-4939-9142-6_12,

such a pipeline for *Saccharomyces cerevisiae* metabolomics via gas chromatography-mass spectrometry (GC-MS).

Sample acquisition and preparation specifics will largely depend on the organism to be studied and the analytical instrument to be used. Regardless of organism, a good metabolomics sampling approach should (1) stop metabolism rapidly and with minimal damage to metabolites, (2) allow for separation of cells from growth medium with minimal leakage from the cells, and (3) extract metabolites of interest reliably and with minimal damage to the metabolites. However, there is currently no single method that allows for reliable, quantitative extraction of all cellular metabolites [1–3]. We will present a method for *S. cerevisiae* that employs cold methanol quenching followed by boiling ethanol extraction, which has been demonstrated to be an effective strategy for use with *S. cerevisiae* and numerous other yeasts [4]. Cold methanol quenching has also previously been employed for *Escherichia coli* [2]; however, more recent data suggest that cold ethanol or rapid filtration are preferable sampling methods due to leakage of intracellular metabolites during a methanol quench [5, 6]. These protocols may be found elsewhere.

The instrument used in data acquisition is often largely a practical matter of availability, but if multiple options are available, the instrument may also be chosen for logistical convenience or based on which classes of metabolites are expected to be of greatest interest in the study. Mass spectrometry coupled with gas- or liquid-chromatographic separation is the most common family of analytical techniques for this purpose; NMR spectroscopy is also frequently employed [7]. The focus of this guide will be on gas chromatography-mass spectrometry, but the various available types of liquid chromatography-mass spectrometry (LC-MS) and NMR are similarly reasonable choices. Each technique involves trade-offs that should be considered with respect to the needs of the experiment if multiple options are available. GC-MS offers sensitive detection for a wide array of metabolites, and makes use of "hard" electron-impact ionization which allows greater ability to distinguish between isomers than most standard LC-MS approaches (an ability that, notably, may be key to providing useful information about several metabolites in the same pathway) at the cost of relatively complex sample preparation required to volatilize many of the compounds to be analyzed. LC-MS generally involves simpler sample preparation but lower resolution and sensitivity of analytes at comparable instrument tiers (e.g., based on cost or nominal vs. accurate mass) to GC-MS [8]. NMR is advantageous in that it is nondestructive for samples, but its substantially lower sensitivity relative to mass spectrometry limits its utility in detecting low-abundance metabolites.

Downstream data processing for metabolomics is sufficiently complex to warrant the development of entire computational pipelines, with components of those pipelines dependent on instrument type used for data acquisition, instrument vendor, and experimental design [8, 9]. Continuing with our focus on GC-MS, we will present an example of such a pipeline for nominal mass GC-MS data (though it is easily generalized to other chromatography variants) that makes use of freely available software when possible.

Application of the results of metabolomics analyses to strain optimization for metabolic engineering will ultimately be highly specific to the system under study and the intended engineering result. Here, we present the tools necessary for start-to-finish collection and processing of such data from *S. cerevisiae*; application of those results should begin with comparison of the metabolic "fingerprint" of different strains in the context of the titer or productivity for the target molecule. Changes in metabolites within the primary pathway of engineering interest serve as a barometer of the effectiveness of the system as a whole, but perhaps the most profound potential value in metabolomics data lies in the identification of other pathways' effects on the pathway central to the synthesis of the desired product. This could help enable identification of cofactor or biosynthetic intermediate limitations that may prevent higher titer or productivity. We recommend making use of pathway enrichment analysis tools (which are included in the MetaboAnalyst package [10] discussed in the data processing section) to aid in identifying such pathway cross talk effects and to identify potential targets for pathway manipulation during further iterations of strain development.

2 Materials

This procedure produces both hazardous chemical waste and biohazardous waste; follow all waste disposal regulations carefully.

2.1 Sample Acquisition and Preparation

1. 15 mL conical centrifuge tubes with corresponding tube racks and centrifuge inserts.
2. Refrigerated centrifuge.
3. Microcentrifuge.
4. Insulated container(s) for cold baths in which the 15 mL tube racks will fit, such as foam ice buckets.
5. Low-temperature thermometer.
6. Temperature-controlled water bath.
7. Centrifugal concentrator (e.g., Labconco CentriVap).
8. 2–3 L ethanol for cold bath(s).
9. Liquid nitrogen (for immersing tubes at end of quenching procedure).
10. Dry ice (optional; *see* **Note 1**).

11. HEPES–methanol: 17 mM HEPES (4-(2-hydroxyethyl)-1-piperazineethanesulfonic acid), methanol (*see* **Note 2**).
12. 60/40 wash: 10 mM HEPES, methanol, ultrapure water. (Total solvent composition should be 60% methanol by volume, and 40% water by volume.)
13. HEPES–ethanol: 2 mM HEPES, 75% ethanol in water (*see* **Note 3**).
14. Ultrapure water.

2.2 GC-MS Data Acquisition

1. GC-MS instrument.
2. Autosampler vials with caps and low-volume inserts.
3. MeOx solution: 40 mg/mL methylhydroxyamine hydrochloride (MeOx), pyridine (*see* **Note 4**). Store at 4 °C in flammable-safe storage.
4. *N*-Methyl-*N*-trimethylsilyltrifluoroacetamide (MSTFA) plus 1% 2,2,2-trifluoro-*N*-methyl-*N*-(trimethylsilyl)-acetamide, chlorotrimethylsilane (TMCS) silylation reagent. Store at 4 °C in flammable-safe storage.

2.3 Data Processing Software

1. Automated Mass Spectral Deconvolution and Identification System (AMDIS [11], freely available at http://chemdata.nist.gov/mass-spc/amdis/downloads/), or the instrument manufacturer's peak deconvolution software.
2. Peak alignment software, such as SpectConnect [12] (freely available at http://spectconnect.gatech.edu/) or MetPP [13] (specifically designed for GCxGC-TOF-MS data, freely available at http://metaopen.sourceforge.net).
3. Other data processing software as desired (*see* **Note 5**).

3 Methods

These sample preparation procedures are optimized for sampling of *S. cerevisiae* in suspension culture (*see* **Note 6**); however, the authors have applied this protocol to a dozen other *Ascomycota* yeasts. While not the recommended method, it may also be applied to *E. coli*, though significant metabolite leakage from cells during quenching will be observed.

1. *Time*: From start to finish, this set of procedures should be expected to occur over a minimum of 3–4 days. Suitable pause points (during which samples may be stored at −80 °C for up to several weeks) are after freezing in liquid nitrogen and after ethanol extraction.
2. *Data quality*: It is very important that all sample handling be performed as rapidly and as consistently as possible, at as cold a

temperature as possible. Metabolism happens on short time scales with entire pools of metabolites potentially turning over in a few seconds, meaning that results can be greatly affected by small variations in prequench handling time or pre-extraction temperature excursions. The physical logistics of the quenching and extraction procedures are nontrivial, and should be given careful consideration when determining experimental design (i.e., frequency of sampling and number of samples per time point). It is recommended that researchers new to these procedures familiarize themselves with the process via a small-scale pilot experiment prior to attempting a full-scale experiment.

3. *Safety*: These procedures involve sample manipulation in open baths of liquid at potentially dangerous temperatures. Wear appropriate gloves to prevent frostbite or scalding and do not reach directly into the liquid baths or liquid nitrogen. These procedures also involve the handling of highly hazardous, flammable chemicals (pyridine, MSTFA). Only handle these materials in a chemical fume hood and with appropriate gloves.

3.1 Quenching

At least 1 day prior to the intended sampling period:

1. Prepare 15 mL conical tubes for the intended samples with HEPES–methanol solution (and water, if needed). 9 mL of HEPES–methanol is needed per sample tube. If the intended sample volume is less than 6 mL, add ultrapure water as well so that the total volume with sample added will be 15 mL (*see* **Note 7**). Chill at −80 °C, overnight.
2. Prepare the 60/40 wash solution; 7 mL is needed per sample tube. Chill at −20 °C overnight, then move to −80 °C 2–3 h prior to beginning sampling (*see* **Note 8**).
3. Prechill the ethanol bath(s) and appropriate centrifuge bucket inserts at −80 °C, overnight.

3.1.1 Day of Experiment

1. Ensure that the tabletop centrifuge is chilled to −9 °C or lower (with prechilled bucket inserts added) prior to sampling.
2. Immediately prior to sampling, remove the ethanol bath and appropriate sample tubes from the freezer, place the sample tubes into the bath, and preloosen the tube caps. Draw a small, insulated dewar of liquid nitrogen and place it nearby but out of the way. The ethanol bath needs to remain below −40 °C during use (*see* **Note 1**); monitor via low-temperature thermometer.
3. At the desired time, use a serological pipette to quickly place samples into the appropriate tubes.

4. Centrifuge the samples for 5 min at 2750 ×*g*. Remove the 60/40 wash from the freezer and place into the ethanol bath while the samples are spinning.
5. Remove samples from the centrifuge immediately and place into the ethanol bath. For each sample, pour off and discard the supernatant and resuspend the pellets in 7 mL 60/40 wash solution (*see* **Note 9**). The samples should not be kept out of the bath for more than 1 min at a time during this step.
6. Centrifuge the samples for 5 min at 3500 rpm.
7. For each sample, discard the supernatant, cap the tube securely, and immediately drop into the liquid nitrogen bath.
8. When all the samples are in the liquid nitrogen, carefully remove them using a slotted spoon and store at −80 °C until extraction.

3.2 Metabolite Extraction

1. Preheat the water bath to 90 °C.
2. Place the HEPES–ethanol solution into a beaker with a stir bar, cover tightly with aluminum foil, and heat with stirring until just boiling.
3. Remove the quenched sample tubes from the freezer and set them next to the hot plate.
4. Add 1 mL of the boiling ethanol solution to each sample tube (*see* **Note 10**).
5. Shake by hand or vortex each sample tube for 3 s to resuspend most of the pellet, and place into the water bath for 3 min, vortexing for 3 s every minute (*see* **Note 9**).
6. Chill the samples at −20 °C for 10 min.
7. Using a 1 mL micropipette, transfer the entire contents of each sample tube into a 1.5 mL microcentrifuge tube.
8. Dry the samples completely in a centrifugal concentrator under vacuum at 40 °C (this will take several hours).
9. Thoroughly resuspend the samples in 200 μL ultrapure water, then centrifuge to form a pellet (*see* **Note 9**).
10. Transfer the supernatant into clean microcentrifuge tubes. If pooled quality control sample(s) are desired, remove a set volume from each tube and pool together (*see* **Note 11**).
11. Return the tubes with the supernatant to the centrifugal concentrator at 40 °C until the samples are fully dry. The samples may then be derivatized for immediate GC-MS analysis, or stored at −80 °C. If the samples are stored for later use, place them back into the concentrator for 10–15 min immediately prior to derivatization to ensure no water from condensation contaminates the samples.

3.3 Derivatization and GC-MS Data Acquisition

1. Make sure that the samples are completely dry prior to derivatization (*see* Subheading 3.2, **step 11**). In a chemical fume hood, add 10 μL of MeOx solution to each sample, scraping the tube with the pipette tip to resuspend the dried analytes as thoroughly as possible.
2. Incubate the samples at 30 °C for 90 min (*see* **Note 12**).
3. In a chemical fume hood, add 90 μL of MSTFA plus 1% TMCS reagent to each sample.
4. Incubate the samples at 37 °C for 30 min.
5. During this incubation, prepare the autosampler vials: add low-volume inserts to each, and label appropriately.
6. Centrifuge the samples at 10,000 × *g* or more for 3 min.
7. Return the samples to the fume hood, and transfer 50 μL of each into the appropriate autosampler vial, being careful to avoid any solid matter at the bottom of the tube, as this will clog the instrument's syringe (*see* **Note 13**). Firmly cap each autosampler vial.
8. Load the samples into the autosampler of the GC-MS and begin the instrument run (*see* **Note 14**).

3.4 Data Processing

1. Using AMDIS or the instrument manufacturer's software, carry out peak deconvolution on all data. For AMDIS, default parameters are typically sufficient. Vendor-specific software typically comes with guidelines for selecting key parameters (e.g., expected peak width).
2. Use SpectConnect, MetPP, or other software of choice to conduct peak alignment. Prior to further statistical analysis, it is usually desirable to filter peaks based on reproducibility after peak alignment (*see* **Note 15**). The data can be checked against a mass spectrometry compound library [14–16] at this stage to identify analytes.
3. Conduct any further statistical analyses desired, per the needs of the experiment. Examples of typical analyses include principal components analysis (PCA), partial least squares discriminant analysis (PLS-DA), metabolite pathway enrichment analysis (MPEA), *t*-tests, and analysis of variance (ANOVA).

4 Notes

1. The ethanol bath will need to remain below −40 °C during use. Depending on sampling frequency, it may be desirable to use two ethanol baths, alternating at each time point so that they have more time to chill between uses. Dry ice may also be used to maintain the temperature during sampling, if necessary.

2. This solution is most easily made by preparing the HEPES in a 1 M stock solution in ultrapure water, then diluting into methanol. Make slightly more HEPES–methanol than the exact amount needed for the experiment, as well as extra 60/40 wash.
3. Significant excess is needed due to boil-off during the extraction procedure. This solution is easily made from the same 1 M HEPES in water stock solution, diluted into the ethanol.
4. Mild heating is typically necessary to fully dissolve the MeOx. It is simplest to make the solution in a well-sealed glass bottle and heat gently in an incubator or a water bath at 37 °C, gently swirling occasionally until all of the crystals have dissolved.
5. The authors routinely use the online suite MetaboAnalyst [10] (freely available at http://www.metaboanalyst.ca/) for principal components analysis (PCA) and other statistical analyses of metabolomics data as needed for a given experiment. XCMS Online [17] (available at https://xcmsonline.scripps.edu/) is another comprehensive metabolomics analysis platform, developed for LC-MS data.
6. Culturing procedure and sampling frequency, time, and cell density will vary according to the needs of the experiment. Cultures should typically be inoculated with as similar a number of cells as possible; this is best accomplished by using an automated cell counter with the seed cultures prior to inoculation of the experimental cultures. Cell density for sampling should be preoptimized based on the sensitivity of the instrument; this can be easily done as part of an initial "training" experiment, as perfect sampling technique and speed is not necessary to simply establish appropriate sampling density. A suggested starting point for sampling is 5 mL samples at 10^7 to 10^8 cells/mL. It may be desirable to take replicate samples to archive for repeated processing and/or RNA collection; if so, consider this when determining the total number of tubes. The replicates may be stored at the quenched stage at −80 °C until needed.
7. As the tubes will be partially immersed in ethanol during the sampling procedure, care should be taken to label them in such a manner that the label will not come off in the ethanol bath. Ethanol-insoluble ink may be used, or the tubes may be labeled near the top (but not on the lids, because dropped lids are common in these time-intensive and low-temperature processes such that spare lids will often be needed). Also, if a very small volume of sample is expected such that the quenching tube has a large volume of water when it is stored overnight, those tubes may need to be removed from the −80 °C freezer early for thawing in case they have frozen (*see* **Note 8**).

8. The 60/40 wash will freeze if kept overnight, or generally kept too long, at −80 °C. This is also a potential issue during the wash step: if the ethanol bath is kept on dry ice and the 60/40 wash is kept in that bath, it may begin to freeze and this should be avoided by rotating out multiple bottles of 60/40 wash.
9. Resuspending the pellets at various stage of the process is one of the more difficult and more variable steps in the process. When resuspending for a 60/40 wash step, using a wooden applicator (commonly used for streaking plates) to help break up the pellet in between vortexings will help resuspend most pellets quite easily, though use of a 1000 μL pipette is also possible as long as one ensures that minimal cell mass gets entrained in the tip. When resuspending after the first centrifugal evaporation, a smaller pipette tip is often necessary to resuspend the pellet effectively. During the boiling ethanol step, a wooden applicator or pipette tip may be used to aid in resuspension, but take care not to keep the tube outside of the water bath for too long.
10. As previously noted, significant excess of this ethanol mixture is necessary due to losses during boiling and while sampling. If the aluminum foil is not immediately replaced on the beaker after some of the mixture is removed for sampling, or if the ethanol mixture is left boiling for too long such that there are substantial evaporative losses, the heated ethanol mixture should be discarded and a new one used, as the former may have too low a concentration of ethanol to allow effective extraction.
11. A significant challenge in metabolomics data collection and processing is minimizing "batch effects"—even when all samples are handled in a very precise manner, GC-MS exhibits "instrument drift" resulting in systematic error introduced into an experiment that may become problematic if the samples are to be run in multiple batches. However, as the derivatization reaction itself will continue while samples are queued to be run, preparing and running the samples in batches is often necessary. (We recommend batches that will take no more than 12 h of total time to run, but smaller batches than this may be necessary.) If the samples are to be run in multiple batches, identical quality control (QC) samples should be prepared from pooled fractions of all the samples in the experiment (Subheading 3.2, **step 10**), and a QC sample run multiple times in each batch. This allows for batch error correction in the data processing steps via locally weighted regression [18].
12. The incubations in the derivatization procedure are ideally done in a heating block with agitation, but a simple heating

block or water bath (with the tubes sealed very carefully and dried with a towel before reopening) may also be used.

13. At this stage, retention index standard solution may be spiked into the samples, if desired [16, 19].
14. As the specifics of the GC-MS operation and method will vary widely according to both the instrument being used and the parameters of the given experiment, a detailed explanation of GC-MS operation is beyond the scope of this text. However, a universally important factor in metabolomics data acquisition is the quality of separation of the compounds prior to reaching the mass spectrometer, and one of the primary ways by which this is controlled is by setting the speed of the oven temperature ramp. Typical protocols for analysis of simple mixtures often specify a ramp rate of 30 °C per minute or faster, and this is seldom appropriate for metabolomics data, as the sample matrices of metabolomics samples are very complex. An appropriate starting point for most yeast metabolomics experiments is 10–20 °C per minute, from which further method optimization can be carried out as desired. (Varying temperature ramp rates may be employed in order to ensure good separation in particularly dense regions of the resulting chromatogram while optimizing for speed in regions where analytes are sparser, for example.) Also, for the purposes of data correction using the quality control samples, a QC sample should be run every 4–8 injections, with two QC injections at the beginning of the batch and two at the end of the batch.
15. The authors have previously [20] removed peaks with 50% or more missing values and 50% or higher coefficient of variation in the quality control samples in a serum-based metabolomics study. Filtering parameters should be set as appropriate to the data quality and needs of the experiment [9].

References

1. Villas-Boas SG, Hojer-Pedersen J, Akesson M, Smedsgaard J, Nielsen J (2005) Global metabolite analysis of yeast: evaluation of sample preparation methods. Yeast 22(14):1155–1169. https://doi.org/10.1002/yea.1308
2. Winder CL, Dunn WB, Schuler S, Broadhurst D, Jarvis R, Stephens GM, Goodacre R (2008) Global metabolic profiling of *Escherichia coli* cultures: an evaluation of methods for quenching and extraction of intracellular metabolites. Anal Chem 80(8):2939–2948. https://doi.org/10.1021/ac7023409
3. Maharjan RP, Ferenci T (2003) Global metabolite analysis: the influence of extraction methodology on metabolome profiles of *Escherichia coli*. Anal Biochem 313(1):145–154
4. Loret MO, Pedersen L, Francois J (2007) Revised procedures for yeast metabolites extraction: application to a glucose pulse to carbon-limited yeast cultures, which reveals a transient activation of the purine salvage pathway. Yeast 24(1):47–60. https://doi.org/10.1002/yea.1435
5. Bordag N, Janakiraman V, Nachtigall J, Maldonado SG, Bethan B, Laine JP, Fux E (2016) Fast filtration of bacterial or mammalian suspension cell cultures for optimal metabolomics

results. PLoS One 11(7):e0159389. https://doi.org/10.1371/journal.pone.0159389

6. McCloskey D, Utrilla J, Naviaux RK, Palsson BO, Feist AM (2015) Fast Swinnex filtration (FSF): a fast and robust sampling and extraction method suitable for metabolomics analysis of cultures grown in complex media. Metabolomics 11(1):198–209. https://doi.org/10.1007/s11306-014-0686-2
7. Larive CK, Barding GA, Dinges MM (2015) NMR spectroscopy for metabolomics and metabolic profiling. Anal Chem 87(1):133–146. https://doi.org/10.1021/ac504075g
8. Dettmer K, Aronov PA, Hammock BD (2007) Mass spectrometry-based metabolomics. Mass Spectrom Rev 26(1):51–78. https://doi.org/10.1002/mas.20108
9. Dunn WB, Broadhurst D, Begley P, Zelena E, Francis-McIntyre S, Anderson N, Brown M, Knowles JD, Halsall A, Haselden JN, Nicholls AW, Wilson ID, Kell DB, Goodacre R, HSMH C (2011) Procedures for large-scale metabolic profiling of serum and plasma using gas chromatography and liquid chromatography coupled to mass spectrometry. Nat Protoc 6 (7):1060–1083. https://doi.org/10.1038/nprot.2011.335
10. Xia JG, Mandal R, Sinelnikov IV, Broadhurst D, Wishart DS (2012) MetaboAnalyst 2.0-a comprehensive server for metabolomic data analysis. Nucleic Acids Res 40(W1): W127–W133. https://doi.org/10.1093/nar/gks374
11. Stein SE (1999) An integrated method for spectrum extraction and compound identification from gas chromatography/mass spectrometry data. J Am Soc Mass Spectrom 10 (8):770–781. https://doi.org/10.1016/s1044-0305(99)00047-1
12. Styczynski MP, Moxley JF, Tong LV, Walther JL, Jensen KL, Stephanopoulos GN (2007) Systematic identification of conserved metabolites in GC/MS data for metabolomics and biomarker discovery. Anal Chem 79 (3):966–973. https://doi.org/10.1021/ac0614846
13. Wei XL, Shi X, Koo I, Kim S, Schmidt RH, Arteel GE, Watson WH, McClain C, Zhang X (2013) MetPP: a computational platform for comprehensive two-dimensional gas chromatography time-of-flight mass spectrometry-based metabolomics. Bioinformatics 29 (14):1786–1792. https://doi.org/10.1093/bioinformatics/btt275
14. Schauer N, Steinhauser D, Strelkov S, Schomburg D, Allison G, Moritz T, Lundgren K, Roessner-Tunali U, Forbes MG, Willmitzer L, Fernie AR, Kopka J (2005) GC-MS libraries for the rapid identification of metabolites in complex biological samples. FEBS Lett 579(6):1332–1337. https://doi.org/10.1016/j.febslet.2005.01.029
15. Ausloos P, Clifton CL, Lias SG, Mikaya AI, Stein SE, Tchekhovskoi DV, Sparkman OD, Zaikin V, Zhu D (1999) The critical evaluation of a comprehensive mass spectral library. J Am Soc Mass Spectrom 10(4):287–299. https://doi.org/10.1016/S1044-0305(98)00159-7
16. Kind T, Wohlgemuth G, Lee DY, Lu Y, Palazoglu M, Shahbaz S, Fiehn O (2009) FiehnLib: mass spectral and retention index libraries for metabolomics based on Quadrupole and time-of-flight gas chromatography/mass spectrometry. Anal Chem 81 (24):10038–10048. https://doi.org/10.1021/ac9019522
17. Tautenhahn R, Patti GJ, Rinehart D, Siuzdak G (2012) XCMS online: a web-based platform to process untargeted Metabolomic data. Anal Chem 84(11):5035–5039. https://doi.org/10.1021/ac300698c
18. Cleveland WS (1979) Robust locally weighted regression and smoothing scatterplots. J Am Stat Assoc 74(368):829–836. https://doi.org/10.2307/2286407
19. Rostad CE, Pereira WE (1986) Kovats and Lee retention indexes determined by gas chromatography/mass spectrometry for organic-compounds of environmental interest. J High Res Chromatog 9(6):328–334. https://doi.org/10.1002/jhrc.1240090603
20. Cipriano RC, Smith ML, Vermeersch KA, Dove ADM, Styczynski MP (2015) Differential metabolite levels in response to spawning-induced inappetence in Atlantic salmon *Salmo salar*. Comp Biochem Phys D 13:52–59. https://doi.org/10.1016/j.cbd.2015.01.001

Chapter 13

Targeted Mass Spectrometry-Based Proteomics Tools for Strain Optimization

Hsien-Chung Tseng and Christine Nicole S. Santos

Abstract

The goal of strain optimization is to create high-performance strains producing compounds of interest at a high titer, yield, and volumetric productivity. The effectiveness of strain optimization relies on methodologies used to aid optimization of native or novel pathways within cells. Many different factors, including mRNA abundance, protein abundance, and enzyme activity/stability, will contribute to the strain performance, which is not often evident by simply monitoring product titers. To this end, targeted proteomics tools utilizing LC-MS-MS (liquid chromatography coupled with tandem mass spectrometry) have recently been developed and can monitor protein levels at great sensitivities. Here, we describe all relevant aspects when developing proteomics tools for strain optimization.

Key words Proteomics, Omics, Strain optimization, Metabolic engineering, Industrial biotechnology, Chromatography, Mass spectrometry

1 Introduction

Proteins are vital parts of living organisms as they are not only the major components for cellular structures but catalytic enzymes in metabolic pathways for cellular functions [1]. To optimize strain performance, synthetic biologists and metabolic engineers have been involved in developing vast arrays of standardized biological genetic tools to control protein expression in an effort to manipulate cellular metabolism. Such genetic tools include but not limited to codon usage, mRNA secondary structure, promoter strength, ribosome binding site strength, terminator strength, mRNA-ribosome occupancy, protein stability, the position of genes in the operon, and the spacing between genetic elements, all of which often profoundly affect the amount of protein expressed [2]. Therefore, a quick, reliable, and multiplexed measurement of expressed proteins from different engineered strains can provide extremely valuable information for iterative strain optimization. Proteomics

Christine Nicole S. Santos and Parayil Kumaran Ajikumar (eds.), *Microbial Metabolic Engineering: Methods and Protocols*, Methods in Molecular Biology, vol. 1927, https://doi.org/10.1007/978-1-4939-9142-6_13,

analysis using LC-MS-MS is the ideal and accurate tool to fulfill this need [3].

Proteomics is typically divided into two categories: discovery and targeted proteomics. Discovery proteomics optimizes protein identification by spending more time and effort per sample in order to maximize the number of detected proteins. In contrast, targeted proteomics strategies limit the number of monitored proteins to achieve the highest sensitivity and throughput for hundreds of samples. During the past few years, targeted proteomics has emerged as a valuable complement to the more widely used discovery proteomics. In fact, targeted proteomics has recently been selected as the Nature Method of the Year for 2012 [4] and is recognized as the most sensitive and specific mass spectrometry-based analysis method to detect and quantify preselected components in a complex sample matrix, for example, the proteolytic digest of the total protein extracted from recombinant *E. coli* cells.

The typical targeted proteomics workflow includes total protein extraction [5], trypsin digestion, and LC-MS-MS data acquisition (Fig. 1). Relative and absolute quantification of proteins are two common approaches for data analysis. The former does not require synthetic peptides; however, the latter utilizes known amount of synthetic (light or isotope-labeled) peptides for calibration. We routinely perform relative quantification by comparing various engineered strains to a control strain without using synthetic peptides [6]. The quantitation of each targeted protein is based on the sum of ion intensity of their three precursor peptides with BSA (bovine serum albumin) as the internal standard to correct for efficiencies of trypsin digestion and MS ionization as well as injection errors. A minimum of cells equivalent to $mL \times OD600 = 3$ is needed. This protocol details the experimental steps for sample preparation and also introduces the Skyline software for method development and data analysis [7].

2 Materials

2.1 Reagents

1. Methanol (LC-MS Grade).
2. Chloroform (HPLC Grade).
3. Water (LC-MS Grade).
4. 1 M formic acid (LC-MS Grade).
5. Sequencing Grade Modified Trypsin (Promega; V5111).
6. Bovine serum albumin BSA (20 mg/mL stock) (NEB; B9000S).

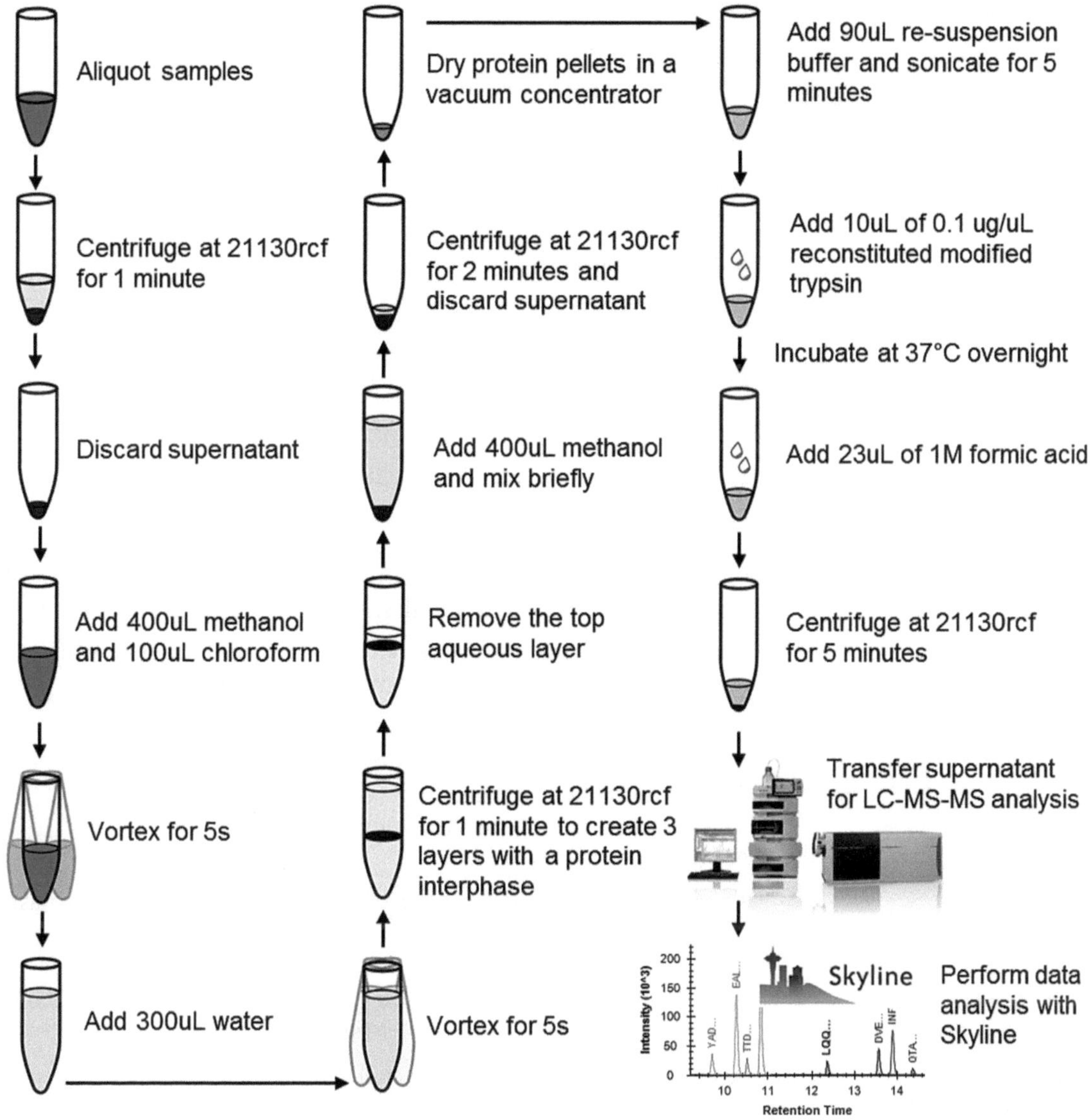

Fig. 1 Overview of procedures for proteomics sample preparation using methanol–chloroform–water precipitation and data analysis using Skyline

2.2 Solutions and Buffers

1. 50 mM ammonium bicarbonate in 10% acetonitrile with 0.1 μg/μL BSA and 1 mM $CaCl_2$ (resuspension buffer) (LC-MS Grade).
2. Mobile phase A: water–acetonitrile 95:5 (v/v) with 0.1% formic acid (LC-MS Grade).
3. Mobile phase B: acetonitrile–water 95:5 (v/v) with 0.1% formic acid (LC-MS Grade).

2.3 Equipment

1. Repeater pipettes or pipettes.
2. Vortex mixer.
3. Advanced Multi-Tube Vortexer.
4. Spectrophotometer.
5. Aspirator.
6. Centrifuge.
7. Ultrasonic bath.
8. LC-MS-MS system (e.g., Agilent 1200 series LC system + 6460 Triple Quad equipped with an electrospray ion source).
9. HPLC column (Agilent Zorbax Eclipse Plus C18 2.1 × 150 mm).

3 Methods

3.1 Cell Harvesting

1. Cultivate cells in defined or complex liquid media in multiwell plates, culture tubes, shake flasks, or fermenters (*see* **Note 1**).
2. Measure cell concentrations with an optical density at 600 nm (OD600) using spectrophotometer (*see* **Note 2**).
3. Equivalent to mL × OD600 = 3 cells (e.g., 0.5 mL of cell broth at OD600 = 6) is transferred to 1.5 mL Eppendorf tubes, and then pelleted by centrifugation (21,130 rcf or 15,000 rpm for Eppendorf centrifuge 5424 for 1 min) (*see* **Note 3**).
4. Discard the supernatant; (*Optional*) the cell pellets can be flash frozen in liquid nitrogen and stored at −80 °C until use.

3.2 Protein Extraction by Chloroform--Methanol–Water Precipitation

1. Add 400 μL methanol and 100 μL chloroform to the tube and vortex for 5 s thoroughly (*see* **Note 4**).
2. Add 300 μL water (mixture will become cloudy with precipitate) and vortex for 5 s (*see* **Note 5**).
3. Centrifuge (21,130 rcf or 15,000 rpm for Eppendorf centrifuge 5424 for 1 min) to generate three layers: a large aqueous layer on top, a circular flake of protein in the interphase, and a smaller chloroform layer at the bottom.
4. Remove the top aqueous layer carefully by using either an aspirator or a pipette, trying not to disturb the protein flake. It is okay to leave some residual aqueous layer (30–50 μL).
5. Add 400 μL methanol (*see* **Note 6**).
6. Vortex for 5 s (*see* **Note 7**).
7. Centrifuge (21,130 rcf or 15,000 rpm for Eppendorf centrifuge 5424 for 3 min), which will slam the dandruffy precipitate against the tube wall.

8. Remove as much methanol as possible by using either an aspirator or a pipette. Be careful, because the pellet is delicate. It is okay to leave some residual methanol (30–50 μL).
9. Dry the protein pellets in a vacuum concentrator (Eppendorf Vacufuge; *Mode*: V-AL (vacuum-alcohol); *Temp*: 30 °C) for 30 min (*see* **Note 8**).
10. Add 90 μL of resuspension buffer.
11. Redissolve protein aggregates by sonication in a sonic bath for 5 min (*see* **Note 9**).
12. (*Optional*) Determine protein concentration with the Bradford Assay according to manufacturer's instructions (*see* **Note 10**).

3.3 Trypsin Digestion

1. Resuspend sequencing grade modified trypsin powder (stored at −20 °C) in 200 μL trypsin resuspension buffer (provided along with the trypsin). The final concentration is 0.1 μg/μL (*see* **Note 11**).
2. Heat the solution at 30 °C for 15 min before use (*see* **Note 12**).
3. Add 10 μL of 0.1 μg/μL reconstituted trypsin (the final trypsin–protein ratio of will be between 1:100 and 1:20 (w/w)) (*see* **Note 13**).
4. Incubate at 37 °C for overnight but not longer than 24 h (between 6 and 24 h).
5. Terminate the reaction by the addition of 23 μL of 1 M formic acid (Trypsin will be inactivated at pH < 4; check the pH with a pH paper if needed) (*see* **Note 14**).
6. Centrifuge (21,130 rcf or 15,000 rpm for Eppendorf centrifuge 5424 for 5 min) to settle down any particulates (*see* **Note 15**).
7. Transfer 80 μL of the supernatant to an LC vial containing an insert or 96-well PCR plate.
8. The samples can then analyzed using LC-MS/MS with an injection volume of 6 μL.

3.4 LC-MS-MS Analysis

1. All samples are analyzed using an LC-MS-MS system consisting of an Agilent 1200 Series liquid chromatograph coupled to an Agilent 6460 Triple Quad equipped with an electrospray ion source with a recommended HPLC setting (Table 1), MS setting (Table 2) and MRM transitions (Table 3) (*see* **Note 16**).
2. The acquired data is then imported to Skyline, which is open source and freely available for academic and commercial use. Skyline can be used for targeted proteomics method creation (Fig. 2) and for data analysis from not only targeted proteomics experiments performed using MRM but the raw data files of full-scan mass spectrometers, such as ion trap and Q-TOF

Table 1
Mobile phase gradients in reversed-phase HPLC

	Time	Mobile phase A	Mobile phase B
1	3.00 min	100%	0%
2	26.50 min	50%	50%
3	29.00 min	100%	0%
4	30.00 min	100%	0%

Table 2
Source parameters for Agilent 6460 QQQ

Source parameter	Value (+)	Value (−)
Gas Temp (°C)	325	325
Gas Flow (L/min)	12	12
Nebulizer (psi)	45	45
SheathGasHeater	350	350
SheathGasFlow	11	11
Capillary (V)	4500	4000
VCharging	1100	500

Table 3
An example of MRM transitions for monitoring *E. coli* DXS (1-deoxy-d-xylulose 5-phosphate synthase) enzyme with BSA as the internal standard

Proteins	Precursor peptides	Product ions	Fragmentor voltage (V)	Collision Energy (V)
EcDxs	YPTLALVDSTQELR	y10, y9, y8, y7	130	25.4
	LPVLFAIDR	y7, y6, y5, y4	130	11.1
	GNAVGVELTPLEK	y9, y7, y5, y4	130	18.3
BSA	LVNELTEFAK	y8, y7, y6, y5	130	14.1
	LGEYGFQNALIVR	y9, y7, y6, y3	130	10.3
	QTALVELLK	y7, y6, y5, y3	130	22.2

instruments generated from various mass spectrometer vendors. A few useful links are summarized here for in-depth Skyline learning of method creation and data analysis (Table 4) (*see* **Note 17**).

3. The quantitation of each targeted protein is based on up to 12 data points (3 precursor peptides × 4 daughter ions) with

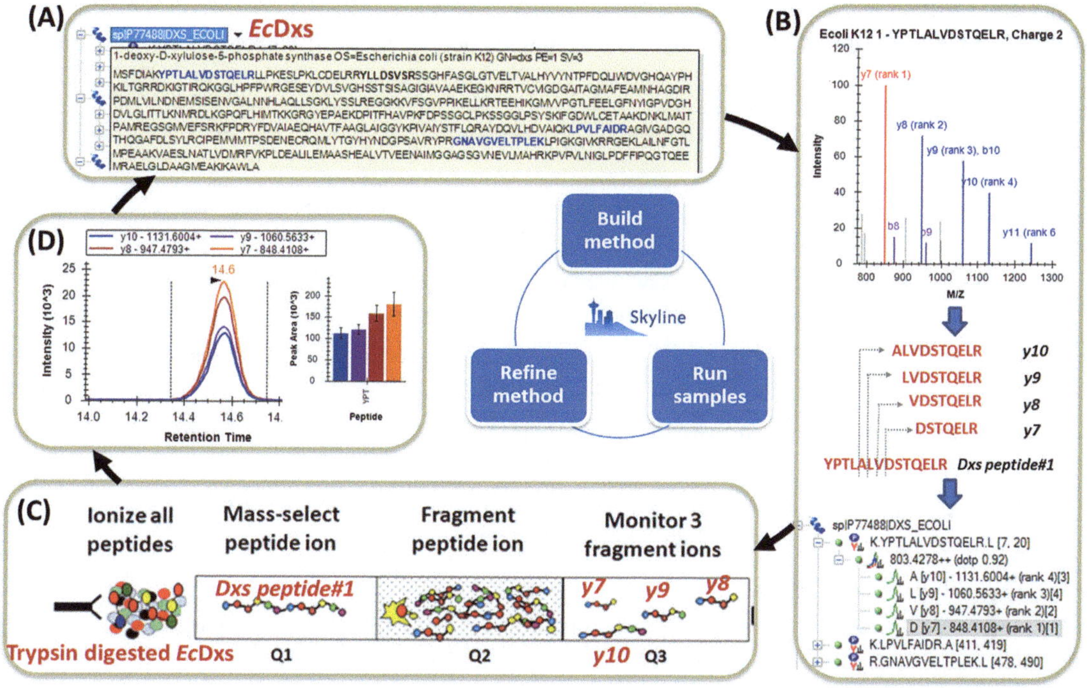

Fig. 2 Screenshot showing method development cycle with Skyline (**a**) The amino acid sequences of proteins of interest are imported into Skyline. An in silico digestion is performed based on user-definded settings; (**b**) Skyline quickly generates MRM transition lists based on available spectral libraries; (**c**) Trypsin-digested samples are injected to LC-MS-MS to validate the MRM transition lists (**d**) Review data and narrow down the transition list (three precusor peptides with four daughter ions per protein) based on the total ion intensity, peak shape, and peptide stability over time

BSA (bovine serum albumin) as the internal standard to correct for efficiencies of trypsin digestion and MS ionization as well as injection errors (Fig. 3).

4. Data can be exported as the csv (comma-separated values) or tsv (Tab Separated Values) type for further analysis with Excel or other software.

4 Notes

1. As the levels of expressed proteins differ at various stages of growth, it is recommended to do time-course sampling to get a more comprehensive view on temporal protein expression and stability. As cell density is low in early- and mid-exponential phases, more culture volume is needed for early time point sampling. However, for a quick proteomics study on comparison of various engineered strains, end time point sampling of cell cultures may be sufficient to provide enough information to guide strain optimization.

Table 4
Links to detailed documentation of Skyline user guides and quick-start videos

Skyline tutorials	Links
Introductory	
Targeted Method Editing	https://skyline.ms/wiki/home/software/Skyline/page.view?name=tutorial_method_edit
Targeted Method Refinement	https://skyline.ms/wiki/home/software/Skyline/page.view?name=tutorial_method_refine
Grouped Study Data Processing	https://skyline.ms/wiki/home/software/Skyline/page.view?name=tutorial_grouped
Existing and Quantitative Experiments	https://skyline.ms/wiki/home/software/Skyline/page.view?name=tutorial_existing_quant
Small Molecule Targets	https://skyline.ms/wiki/home/software/Skyline/page.view?name=tutorial_small_molecule
Full-Scan Acquisition Data	
MS1 Full-Scan Filtering	https://skyline.ms/wiki/home/software/Skyline/page.view?name=tutorial_ms1_filtering
Targeted MS/MS	https://skyline.ms/wiki/home/software/Skyline/page.view?name=tutorial_targeted_msms
Data Independent Acquisition	https://skyline.ms/wiki/home/software/Skyline/page.view?name=tutorial_dia
Advanced Topics	
Absolute Quantification	https://skyline.ms/wiki/home/software/Skyline/page.view?name=tutorial_absolute_quant
Custom and Live Reports	https://skyline.ms/wiki/home/software/Skyline/page.view?name=tutorial_custom_reports
Advanced Peak Picking Models	https://skyline.ms/wiki/home/software/Skyline/page.view?name=tutorial_peak_picking
iRT Retention Time Prediction	https://skyline.ms/wiki/home/software/Skyline/page.view?name=tutorial_irt
Collision Energy Optimization	https://skyline.ms/wiki/home/software/Skyline/page.view?name=tutorial_optimize_ce
Spectral Library Explorer	https://skyline.ms/wiki/home/software/Skyline/page.view?name=tutorial_library_explorer
Quick-start videos	
Method Editing	https://skyline.ms/wiki/home/software/Skyline/page.view?name=video_0-2
Results Analysis	https://skyline.ms/wiki/home/software/Skyline/page.view?name=video_0-5
Existing and Quantitative Experiments	https://skyline.ms/wiki/home/software/Skyline/page.view?name=video_0-5b

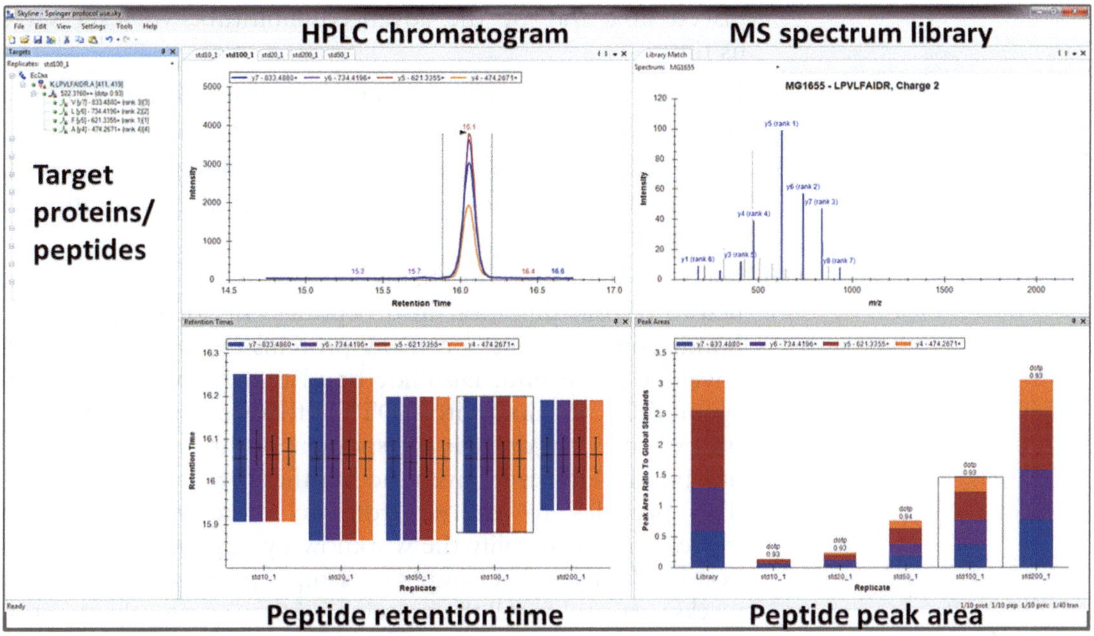

Fig. 3 Screenshot showing data analysis with Skyline

2. Accurate OD600 measurement is very crucial as it is used for data normalization. Peptides of some housekeeping proteins (e.g., RpsA) or enzymes (e.g., GapA) in central metabolism may be used as the reference standards, inferring abnormality in possible sampling errors. If needed, they might also be used for data correction from sample preparation errors. An experimental study with at least triplicate is recommended to have enough statistical power to make observations meaningful.
3. For a small experiment with only a handful of samples, you can measure cell densities of all samples, and aliquot volume based on the actual OD600. However, adjusting pipetting volume for each sampling takes time, especially for a large experiment performed on 96dwp. Alternatively, you can measure cell densities of just a few samples and aliquot the same volume for all samples. The final expression level can later be corrected by the actual OD600 of each sample.
4. Pipet to break the chunk of cell pellet if necessary.
5. We find that using a repeater pipette actually saves a lot of time for sample preparation.
6. Check if the dry flake is in the bottom of tube. It is critical to mix well the methanol and chloroform; otherwise, the flake will not settle to the bottom after centrifugation.
7. If there are many samples to prepare, we find that, instead of vortexing individual tubes, mixing the samples (e.g., 48 tubes)

placed in a tray by the advanced multitube vortexer saves us time.

8. If there is no vacufuge to use, samples can also be dried out in a chemical hood at room temperature, but it will take longer.
9. After sonication, white emulsified samples should be observed. It is normal to see some precipitates ever after sonication as they are cell debris.
10. It is optional to further chemically modify the protein extract before trypsin digestion into peptides. In bacteria, disulfide bonds confer stability on many proteins exported to the cell envelope or beyond. Disulfide bond reduction by a chemical agent, (e.g., dithiothreitol (DTT)) followed by alkylation of the cysteines can prevent the spontaneous formation of disulfide bonds. However, given the fact that there are many candidate peptides to choose from to monitor one protein expression, we simplify the workflow by skipping the disulfide bond reduction. The simplified sample preparation workflow works very well for all the proteins (~50) we currently monitor.
11. Store the reconstituted trypsin enzymes at −80 °C until use. Thaw the reconstituted trypsin at room temperature and place on ice immediately after thawing. Remove the amount of trypsin needed, and then refreeze the unused portion. To maintain maximum product activity, limit the number of freeze-thaw cycles to five or dispense into single-use aliquots after resuspending.
12. Trypsin specifically hydrolyzes peptide bonds at the carboxyl side of lysine and arginine residues. Sequencing grade modified trypsin is provided in 20 μg aliquots with a stability-optimized resuspension buffer. A protease–protein ratio of 1:100 to 1:20 (w/w) is recommended by the manufacturer.
13. This protocol was established based on recombinant *E. coli* strains with plasmid-based or chromosomal expression of proteins. If samples from different expression context are used, it is required to perform **step 11** in Subheading 3.2 to guide tuning the trypsin–protein ratio.
14. To avoid exposing to concentrated formic acid, use the chemical hood when pipetting.
15. Solid waste materials can be disposed in the white biowaste bins; all liquid waste needs to be collected and treated as the aqueous chemical waste; HPLC sample vials should be collected in a jar and treated as the solid chemical waste.
16. This protocol is not specific to a particular sample type or set of proteins, and it is possible to monitor as many proteins as possible. However, there is a limit in transitions per chromatographic run due to noise and sensitivity. For our LC-

MS-MS system, we include up to 240 transitions per chromatographic run, equivalent to 20 monitored proteins (three precursor peptides with four daughter ions per protein) or 60 proteins (one precursor peptide with four daughter ions per protein). It is fine to use just one peptide as a surrogate for the protein of interest if multiplexing is the goal. To monitor more than 60 proteins, a second injection with a method containing the remaining transitions is needed.

17. Minor modifications in source parameters and MRM transitions are required when using mass spectrometer instruments from other vendors (AB SCIEX, Thermo-Scientific, Waters, etc.) [7].

Acknowledgments

This work was supported by Manus Bio (http://www.manusbio.com).

References

1. Altelaar a FM, Munoz J, Heck AJR (2013) Next-generation proteomics: towards an integrative view of proteome dynamics. Nat Rev Genet 14:35–48
2. Edition I (2016) Synthetic biology–the synthesis of biology. Angew Chem Int Ed Engl 56 (23):6396–6419
3. Batth TS, Singh P, Ramakrishnan VR et al (2014) A targeted proteomics toolkit for High-throughput absolute quantification of Escherichia coli proteins. Metab Eng 26:48–56
4. Marx V (2012) Targeted proteomics. Nat Methods 10:19–22
5. Wessel D, Flügge UI (1984) A method for the quantitative recovery of protein in dilute solution in the presence of detergents and lipids. Anal Biochem 138:141–143
6. Walters B, Giaw C, Sagliani K, Shankar S, Stephanopoulos G (2016) Overcoming heterologous protein interdependency to optimize P450-mediated Taxol precursor synthesis in Escherichia coli. Proc Natl Acad Sci USA 113 (12):3209–3214. https://doi.org/10.1073/pnas.1515826113
7. Egertson JD, MacLean B, Johnson R, Xuan Y, MacCoss MJ (2015) Multiplexed peptide analysis using data-independent acquisition and Skyline. Nat Protoc 10:887–903

Chapter 14

A Protocol for the Construction and Curation of Genome-Scale Integrated Metabolic and Regulatory Network Models

Sriram Chandrasekaran

Abstract

Genome-scale metabolic network models have been widely used over the last decade and have been shown to successfully predict the metabolic behavior of many organisms. Yet the complexity of metabolic regulation often limits the accuracy of these models. Integrative modeling approaches have recently been developed that combine metabolic and regulatory networks, thereby expanding the capabilities and accuracy of genome-scale modeling. This chapter provides a guide to reconstruct and curate such integrated network models. Specifically, this protocol describes the PROM (Probabilistic Regulation of Metabolism) and GEMINI (Gene Expression and Metabolism Integrated for Network Inference) approaches. PROM is an automated method for the construction of integrated metabolic and transcriptional regulatory network models, while the GEMINI approach curates the integrated network models using transcriptomics and phenomics data. GEMINI represents the first attempt at applying well-established curation tools that exist for metabolic networks to be applied for curating regulatory networks. The integrated network models generated by these approaches enable the mechanistic integration of diverse biological data and can identify novel strategies to engineer cellular metabolism.

Key words Systems biology, Metabolic networks, Transcriptional regulatory networks, Genome-scale, Modeling

1 Introduction

Cellular metabolism is a complex network comprising hundreds of metabolites, enzymes, and regulatory factors. The size of the metabolic network and the complexity of regulation makes engineering these systems a daunting task. In addition, there are numerous ways in which the network can be rewired to achieve a specific goal for metabolic engineering. Choosing the most optimal strategy among the staggering number of possible solutions is another significant challenge.

Christine Nicole S. Santos and Parayil Kumaran Ajikumar (eds.), *Microbial Metabolic Engineering: Methods and Protocols*, Methods in Molecular Biology, vol. 1927, https://doi.org/10.1007/978-1-4939-9142-6_14,

Computational models have been developed recently to address this complexity and systematically identify strategies for engineering cell metabolism. One of the most popular computational tools for metabolic engineering applications is constraint-based modeling (CBM) [1]. In this approach, a stoichiometric model of metabolism is built by connecting the metabolites with the reactions they participate in, and subsequently representing this network in the form of a matrix. This static representation of the network can be simulated using genome-scale modeling tools. The most common approach—called flux balance analysis (FBA)—uses linear programming to identify an optimal set of reaction fluxes that maximizes a specific objective (usually biomass formation or metabolite of interest) [2].

Genome-scale models have been constructed for many industrially relevant model systems [3]. These models have been applied successfully to design engineered strains. The predictions from these models involve genetic manipulations such as addition or removal of reactions from the metabolic network to achieve an engineering objective. For example, a metabolic model of *E. coli* was used to predict a triple gene knockout strategy for improving lycopene production [4]. Similar approach has been used to improve the production of L-valine and L-threonine using *E. coli* [5, 6]. The numerous achievements of genome-scale modeling tools for fundamental biological discovery and engineering are described elsewhere [7].

Despite these successes, one of the fundamental reasons why the predictions from these models often fail to match experimental observations is that these models do not incorporate any information on metabolic regulation [8, 9]. For example, the model microbe *E. coli* has hundreds of transcription factors (TFs) that can turn genes on or off in a combinatorial fashion. In higher organisms the complex network of regulatory processes comprises noncoding RNAs, numerous posttranslational modifications, and posttranscriptional regulation such as alternate-splicing. While metabolism is highly conserved across the phylogenetic tree and the same engineering concepts (flux-, mass-, and energy-balance analysis) are effective in simulating metabolism across all organisms, the complexity of regulation poses unique challenges necessitating the development of novel theoretical tools.

In addition to enhancing the accuracy of the metabolic models, incorporation of regulatory processes can provide additional access point for engineering a cell. Large scale metabolic changes can be orchestrated by altering few key TFs. Further metabolic changes cannot be done without consideration of regulation because cells have evolved complex regulatory mechanisms to maintain homeostasis that can counteract engineered strategies. The focus of metabolic engineering is slowly shifting toward modifying regulatory processes [10]. Metabolic regulation can be indirectly inferred through high-throughput measurements such as transcriptomics or proteomics, which provide a snapshot of cellular activity. These omics data in combination with genome-scale models can be used to identify a subset of reactions that are active in a specific condition [8]. However, the underlying molecular agents that cause these regulatory effects are unknown in these approaches. Further the impact of engineering regulatory factors, such as deletions of transcription factors, cannot be simulated without a mechanistic integrated regulatory-metabolic model.

A pioneering approach that combines regulation with genome-scale metabolic networks is Regulatory-FBA or R-FBA. Covert et al. used R-FBA to enhance the predictive accuracy of genome-scale models [11, 12]. Using R-FBA, they were able to correctly predict the growth outcomes of *E. coli* with 78.7% accuracy in 13,750 different growth conditions involving different nutrient sources, or after deletion of metabolic genes and transcription factors. Over 18% of these outcomes were correctly predicted only when regulation was incorporated in the model.

Nevertheless, R-FBA has several limitations. This approach assumes that genes operate in a switch-like fashion with two states—completely active (ON) or inactive (OFF). In reality, a gene's expression can take on a continuous range of values. As a result, predictions made with this approach are not highly accurate for eukaryotic organisms with complex regulation such as yeast. In addition, the Boolean rules that connect TF activity to target gene activity for R-FBA are written manually; this process can become daunting for larger networks comprising thousands of interactions, which is the case for most organisms.

The *Probabilistic Regulation of Metabolism* (PROM) approach overcomes these limitations by creating a probabilistic model of transcriptional regulation, thereby allowing for continuous values for gene states [13]. In addition, PROM automates the process of quantifying the relationship between transcription factors and target genes, and requires little manual annotation compared to R-FBA. PROM can directly use high-throughput interaction data as input and leads to more comprehensive and accurate integrated models of metabolism and transcriptional regulation. Other types of integrated network models have also been recently developed that combine regulatory mechanisms with metabolic networks [14–18] and their strengths and weaknesses are reviewed elsewhere [18–20].

Despite promising tools like PROM and R-FBA, the use of integrated network models has been limited due to the lack of large scale regulatory interaction data. Further existing interaction data sets are noisy with significant amount of false positive interactions. Covert et al. manually curated the R-FBA model by comparison with transcriptomics data [12]. They identified cases where the Boolean rules for regulation did not match experimentally observed measurements upon perturbation of a TF, and used this data to create new rules and refine incorrect regulatory rules. Given the large size of the regulatory network—on the order of thousands, this manual process is not scalable for most integrated models. While multiple approaches now exist for the construction of integrated models, systematically curating the models is a significant challenge.

The *Gene Expression and Metabolism Integrated for Network Inference* (GEMINI) algorithm addresses this challenge by curating regulatory networks and integrated metabolic-regulatory network models [21]. The algorithm uses the metabolic network model to reconcile inconsistencies across different data types. GEMINI integrates a compendium of candidate regulatory interactions with the metabolic network to predict their systems-level effect on metabolism and growth phenotypes. GEMINI was applied to curate an integrated metabolic-regulatory network model of *S. cerevisiae* involving 25,000 regulatory interactions influencing 1597 metabolic reactions. The integrated network model quantitatively predicted transcription factor knockout phenotypes with significantly higher accuracy after curation by GEMINI. GEMINI has also been shown to be more effective in curating regulatory networks than gene expression-based approaches [21]. It is the only approach that uses orthogonal information from metabolism to curate regulatory networks, and thus it can assess the quality of networks built by other reverse-engineering approaches [21]. GEMINI can iteratively expand the comprehensiveness and accuracy of integrated models by incorporating new regulatory interactions obtained from experimental or computational approaches.

The process of automated curation by GEMINI is conceptually similar to using automated approaches such as GrowMatch to resolve inconsistencies in the metabolic network [22]. The GrowMatch approach utilizes a universal library of metabolic reactions to identify those that resolve an observed inconsistency. Similarly, GEMINI can identify a set of regulatory interactions from high-throughput data that are consistent with an observed phenotype. There are several detailed protocols for constructing and curating metabolic networks [23–25]. The GEMINI approach aims to exploit this resource for also curating regulatory network models. This protocol will discuss the steps involved in construction and curating these integrated metabolic regulatory network models (*see* **Notes 1–4** for additional comments).

1.1 Theory

1. *FBA*—flux balance analysis (FBA) approach calculates the flux of metabolites through various reactions in a metabolic network [2]. There is a large range of possible fluxes for each reaction in a metabolic network. FBA determines an optimal solution that maximizes the flux through an objective function. The cellular objective is usually set to be the model's biomass reaction.

2. *FVA*—FBA outputs one among many possible flux solutions that maximize a given objective function. Flux Variability Analysis (FVA) captures the entire range of possible flux distributions for each reaction at the optimal objective [26].
3. *PROM*—the probabilistic regulation of metabolism (PROM) algorithm enables the automated integration of Transcriptional Regulatory Networks (TRNs) with metabolic networks into a unified quantitative model [13]. PROM takes in metabolic networks, TRNs (inferred from high-throughput data), and gene expression data as inputs, to build an integrated metabolic-regulatory network model that can predict the impact of TFs on the metabolic network.
4. *GEMINI*—the Gene Expression and Metabolism Integrated for Network Inference (GEMINI) algorithm curates integrated metabolic-regulatory network models using gene expression data and known metabolic phenotypes as input [21]. Through an iterative process, GEMINI can progressively incorporate new regulatory interactions, improve the comprehensiveness and accuracy of the integrated metabolic-regulatory networks.
5. *Network Inference*—regulatory network inference refers to the process of identifying regulatory interactions from high-throughput data such as transcriptomics or proteomics using statistical or machine-learning algorithms [19, 27].

2 Materials

2.1 Resources

Please not that this is not an exhaustive list. This list was compiled based on datasets used in recent genome-scale integrated modeling studies.

1. Databases for regulatory networks—TRANSFAC (http://gene-regulation.com), Yeastract (http://yeastract.com), RegulonDB (http://regulondb.ccg.unam.mx/), BioCyc (BioCyc.org).
2. Databases for metabolic networks BIGG (http://bigg.ucsd.edu/), Metabolic Atlas (http://www.metabolicatlas.org/), SEED (http://modelseed.org/), BioCyc (BioCyc.org), Palsson lab (http://systemsbiology.ucsd.edu/InSilicoOrganisms/OtherOrganisms).
3. Database for transcriptomics—GEO (https://www.ncbi.nlm.nih.gov/geo/), M3D (http://m3d.mssm.edu/).

2.2 Software Prerequisites

1. MATLAB 7.0 or above (http://www.mathworks.com).
2. libSBML programming library (http://www.sbml.org).
3. SBMLToolbox version (http://www.sbml.org).

4. The COBRA Toolbox (http://www.cobratoolbox.org).
5. GLPK (http://glpkmex.sourceforge.net) linear programming (LP) solver.
6. Download PROM and GEMINI MATLAB code (http://www.sriramlab.org/software/) and add them to the MATLAB search path.

2.3 User Data

The first step in curation by GEMINI is the construction of an integrated metabolic and transcriptional regulatory network model. The construction of an integrated model using PROM requires the following data sets (**items 1–3**) described below. In addition, curation by GEMINI requires additional phenotype data (**item 4**).

1. Genome-scale metabolic network model for the organism of interest: Metabolic network reconstructions are manually curated resources and models for several organisms are publicly accessible in the databases listed above.
2. Regulatory network: The regulatory network consists of a list of transcription factors and their interactions with target genes. These transcriptional regulatory networks have generally been constructed based on high-throughput protein–DNA interaction data or statistical inference of functional relationships from genomic and transcriptomic data.
3. Transcriptomics data: This consists of a compendium of gene expression profiles measured under different conditions for the organism of interest. This will enable PROM to quantify of the impact of expression changes of transcription factors on their corresponding target genes.
4. Phenotype data: The phenotype data contains the measurement of growth rate, viability, or metabolite secretion on TF deletion strains under different conditions.

3 Methods

3.1 Formatting the Input Data

An example dataset for this tutorial is available at sriramlab.org/software. This has all the required datasets for running GEMINI for the model organism *S. cerevisiae*. Load the MATLAB data file "*yeast_gemini_data.mat*". The following datasets are in the file:

1. *regulator*, *targets*—set of 31,075 interactions between 177 TFs and 863 targets. The regulatory network is in general provided as an array of regulators and matching target genes. For example, interactions between Regulator A with Genes A and B is represented as,

```
>> Regulator = {'Reg A'; ' Reg A'; 'Reg C'}; Targets
={'Gene A';'Gene B';'Gene C'};
```

Note that the names or identifiers used in the regulatory data should match the names/ids used for gene expression data.

2. *expression, expressionid*—transcriptomics data along with their identifiers. An array of identifiers for each row should be included in *expressionid*. The rows of the gene expression data correspond to different transcripts while columns correspond to different conditions. The data should preferably be normalized, imputed and measured across several growth conditions.
3. *model*—model is the metabolic model for the organism (obtained from COBRA toolbox through ReadCbModel command). In this case, the yeast metabolic model is provided with media set to be glucose minimal media. In general, the model should be set to a specific growth condition under study.
4. *v11 v12*—these are the minimum (v11) and maximum (v12) possible flux through each reaction in the model for the given condition. This data is obtained through flux variability analysis (either using FastFVA or *fluxvariability* command in COBRA toolbox),

```
>> [v11,v12] = fluxvariability(model);
```

5. *phenotype*—TF knockout data for comparing the accuracy of the integrated model and identifying TFs that needs curation. This is a binary vector (true/false values) and should be the same length as the number of TFs in the model.
6. *type*—a string describing the knockout phenotype data. It should be either "lethal" (default) or "suboptimal." Lethal implies that the values in the phenotype data represent lethality (1) or nonessential (0).
7. *OPTIMAL_THRESH* is the threshold for determining lethal/nonlethal phenotypes; by default, a knockout that grows less than 5% of the wild type growth rate is considered lethal and less than 95% of the wild-type growth rate is considered suboptimal. This is an optional parameter. By default, OPTIMAL_-THRESH is set at 0.05 for "lethal" type and 0.95 for "suboptimal" data.
8. *subsets*—represents the subset of TFs for which GEMINI should be run. By default, GEMINI curates the network of all TFs.

3.2 Building an Integrated Model Using PROM

Running PROM:

Syntax: growthrate = promv2(model, expression, expressionid, regulator, regulated, litevidence, prob_prior, subsets, fluxupperbound, fluxlowerbound);

```
>> growthrate = promv2(model, expression, expressionid,
regulator, targets, [], [], {'YAL051W'}, v11, v12);
```

Running this command should give a growth rate output of 1.34.

If the interactions between TFs and targets were determined from literature (instead of inference from high-throughput data), the probabilities can be directly provided for running PROM, instead of calculating the probabilities from transcriptomics data. The "*prob_prior*" variable should have the user defined probabilities and should be same length as the interaction data. The corresponding interactions should be flagged as true (value set to 1) in the "*litevidence*" variable. For example, running PROM with the first ten values in the variable "*litevidence*" set to 1 and the rest set to 0 would over-ride the transcriptomics-based calculation of probabilities for the first ten interactions and PROM would use the corresponding probabilities in "*prob_prior*" variable.

If PROM is run without a specified TF name, like "YAL051W" in the above example, then PROM calculates the predicted growth rate for all TF deletions.

```
>> growthrate = promv2(model, expression, expressionid,
regulator, targets, [], [], [], v11, v12);
```

3.3 Resolving Inconsistencies Using GEMINI

Comparison of the predicted growth rate by PROM (the variable *growthrate* in the above example) with experimental growth rate data suggests that the integrated model incorrectly predicted deletion of 7 TFs (YDR096W, YFR034C, YGL073W, YIL131C, YKL185W, YML007W, YMR037C) as lethal; these predictions were not observed experimentally (phenotype score = 0, implying nonessential). To resolve the inconsistency, we run GEMINI for the TF Ash1 (YKL185W) that was predicted incorrectly.

Syntax:

[f,initial_network,final_network] = GEMINI(model,expression,expressionid,regulator,targets,phenotype,type,subsets,v11, v12)

```
>> [f,initial_network,final_network] =
GEMINI(model,expression,expressionid,regulator,targets,phen
otype,'lethal',{'YKL185W'},v11,v12);
```

```
>> [f,initial_network,final_network] =
GEMINI(model,expression,expressionid,regulator,targets,phen
otype,'lethal',{'YKL185W'},v11,v12,[],[],'probability');
```

GEMINI can be run in two different modes—The default version prioritizes interactions for removal based on their impact on metabolic flux. The second version prioritizes interactions based on the interaction probability. The "Probablilty" mode is faster but is not as accurate as the flux-based default mode. The algorithm outputs the refined network and the growth rate (f) after knock out of the specific TF using the curated network. Interactions that potentially lead to the inconsistency are removed by GEMINI. The refined model can be used with PROM to simulate the impact of TF inhibition in various conditions (*see* **Note 5–9**).

The integrated network model curated using GEMINI outperforms the un-curated high-throughput interaction data even in new growth conditions [21]. For example, curation using TF knockout data from glucose minimal media improves predictions in other conditions such as minimal media with galactose as sole carbon source or minimal media with urea as sole nitrogen source. Benchmarking GEMINI curated and un-curated networks with validated "gold standard" interaction datasets suggest that this process leads to a significant improvement in identification of validated interactions (*see* **Note 10–14**) [21].

4 Notes

1. Integrated metabolic-regulatory models expand the scope of genome-scale modeling by incorporating the role of regulatory mechanisms; transcriptional regulation is the most common regulatory process that is integrated with metabolic networks.
2. Metabolic networks are created in a bottom-up manner based on genomic and bibliomic data. Regulatory networks are derived from top-down approaches such as high-throughput omics data and statistical inference tools.
3. Very few integrated network models currently exist because detailed regulatory information is not available for many systems. Even for those with transcriptional regulatory information, these datasets often contain false positives and cannot be used directly.
4. The process of curating regulatory networks isn't well established as metabolic networks. GEMINI represents the first attempt at applying well-established curation tools that exist for metabolic networks to be applied for regulatory networks.
5. Note that GEMINI may take several minutes for each TF (>10 min/TF depending on the network size).
6. Some TFs may not be active in glucose rich conditions but may have an impact on other conditions. Performing this analysis in

multiple conditions can lead to further curation of the regulatory network.

7. Comparison of GEMINI curated interactions across different growth conditions revealed that some interactions led to inconsistencies with experimental data only in certain conditions. Such conditionally curated interactions most likely result from posttranscriptional regulation of TFs through signaling or posttranslational modification. For example, some TFs are activated only in in specific conditions through phosphorylation or other regulatory mechanisms [21].
8. Not all types of inconsistencies can be resolved using GEMINI. GEMINI specifically addresses inconsistencies that involve the model incorrectly predicting an observed growth phenotype (model prediction—not growing or suboptimal growth; experimental data—growth or optimal growth).
9. The opposite scenario (model prediction—growth; experimental observation—no growth) can be resolved by incorporating new regulatory interactions into the model. New interactions can be obtained from experimental data such as Chromatin Immunoprecipitation followed by sequencing (Chip-Seq) or TF knockout followed by transcriptomics profiling. New interactions can also be obtained through statistical inference approaches such as CLR [28], ARACNE [29], or ASTRIX [30]. These interactions can then be added to the integrated network model to assess if the inconsistency is resolved. If new inconsistencies arise due to the addition of new interaction data, they can be refined with another round of curation by GEMINI.
10. Inconsistencies can also arise due to noise in gene expression data. Transcriptomics data used as input should ideally incorporate a wide range of conditions. The purpose of using gene expression data is to quantify the relationship between transcription factors and target genes. Using transcriptomics data under a wide range of conditions provides a robust estimate of this relationship.
11. GEMINI assumes that the observed inconsistencies in the integrated model arise due to errors in the regulatory network. Hence it is essential that the transcriptomics data used as input for GEMINI is of high quality. Similarly, the inconsistencies arising due to the metabolic network should be resolved using approaches like GrowMatch before being used with GEMINI.
12. Note that the growth rate (variable "f," in the above example) output from GEMINI is only semiquantitative unless the network was curated using quantitative growth data.
13. By default, GEMINI curates the regulatory interactions for all the transcription factors in the network, and the predicted

growth rate of TF knockout strains after curation is outputted in alphabetical order of the transcription factors identifiers.

14. PROM and GEMINI predict phenotypes based on a gene's effect on metabolism; they cannot determine correctly the phenotypes of genes with nonmetabolic functions.

In summary, this chapter provides a protocol for integrating transcriptomic, metabolic, and growth phenotype data to construct and curate integrated models of metabolic and regulatory networks. While GEMINI represents an important first step in curating integrated network models, the accuracy of these models still needs to be improved significantly. Further, the repertoire of regulatory processes captured by these models needs to be expanded beyond transcriptional regulation. Comprehensive integrated models of metabolic and regulatory processes can enable the mechanistic integration of omics data at different biological scales, such as genomics, transcriptomics, proteomics, metabolomics, and fluxomics data, and pave the way for whole-cell models of complex organisms [31].

References

1. Price ND, Papin JA, Schilling CH, Palsson BO (2003) Genome-scale microbial in silico models: the constraints-based approach. Trends Biotechnol 21(4):162–169 doi: S0167779903000301
2. Orth JD, Thiele I, Palsson BØ (2010) What is flux balance analysis? Nat Biotechnol 28 (3):245–248
3. King ZA, Lloyd CJ, Feist AM, Palsson BO (2015) Next-generation genome-scale models for metabolic engineering. Curr Opin Biotechnol 35:23–29
4. Alper H, Miyaoku K, Stephanopoulos G (2005) Construction of lycopene-overproducing E. coli strains by combining systematic and combinatorial gene knockout targets. Nat Biotechnol 23(5):612–616
5. Park JH, Lee KH, Kim TY, Lee SY (2007) Metabolic engineering of Escherichia coli for the production of L-valine based on transcriptome analysis and in silico gene knockout simulation. Proc Natl Acad Sci 104 (19):7797–7802
6. Lee KH, Park JH, Kim TY, Kim HU, Lee SY (2007) Systems metabolic engineering of Escherichia coli for l-threonine production. Mol Syst Biol 3(1):149
7. O'Brien EJ, Monk JM, Palsson BO (2015) Using genome-scale models to predict biological capabilities. Cell 161(5):971–987
8. Lewis NE, Nagarajan H, Palsson BO (2012) Constraining the metabolic genotype-phenotype relationship using a phylogeny of in silico methods. Nat Rev Microbiol 10 (4):291–305. https://doi.org/10.1038/nrmicro2737
9. Herrgard MJ, Lee BS, Portnoy V, Palsson BO (2006) Integrated analysis of regulatory and metabolic networks reveals novel regulatory mechanisms in Saccharomyces cerevisiae. Genome Res 16(5):627–635. https://doi.org/10.1101/gr.4083206 gr.4083206
10. Farmer WR, Liao JC (2000) Improving lycopene production in Escherichia coli by engineering metabolic control. Nat Biotechnol 18 (5):533–537
11. Covert MW, Schilling CH, Palsson B (2001) Regulation of gene expression in flux balance models of metabolism. J Theor Biol 213 (1):73–88. https://doi.org/10.1006/jtbi.2001.2405 S0022-5193(01)92405-1
12. Covert MW, Knight EM, Reed JL, Herrgard MJ, Palsson BO (2004) Integrating high-throughput and computational data elucidates bacterial networks. Nature 429(6987):92–96. https://doi.org/10.1038/nature02456 nature02456
13. Chandrasekaran S, Price ND (2010) Probabilistic integrative modeling of genome-scale metabolic and regulatory networks in Escherichia coli and mycobacterium tuberculosis.

Proc Natl Acad Sci U S A 107 (41):17845–17850. https://doi.org/10.1073/pnas.1005139107 1005139107

14. Carrera J, Estrela R, Luo J, Rai N, Tsoukalas A, Tagkopoulos I (2014) An integrative, multiscale, genome-wide model reveals the phenotypic landscape of Escherichia coli. Mol Syst Biol 10(7):735
15. O'brien EJ, Lerman JA, Chang RL, Hyduke DR, Palsson BØ (2013) Genome-scale models of metabolism and gene expression extend and refine growth phenotype prediction. Mol Syst Biol 9(1):693
16. Lee S, Zhang C, Kilicarslan M, Piening BD, Bjornson E, Hallström BM, Groen AK, Ferrannini E, Laakso M, Snyder M (2016) Integrated network analysis reveals an association between plasma mannose levels and insulin resistance. Cell Metab 24(1):172–184
17. Shlomi T, Eisenberg Y, Sharan R, Ruppin E (2007) A genome-scale computational study of the interplay between transcriptional regulation and metabolism. Mol Syst Biol 3:101. https://doi.org/10.1038/msb4100141 msb4100141
18. Imam S, Schäuble S, Brooks AN, Baliga NS, Price ND (2015) Data-driven integration of genome-scale regulatory and metabolic network models. Front Microbiol 6:409
19. Chandrasekaran S (2014) Predicting phenotype from genotype through reconstruction and integrative modeling of metabolic and regulatory networks. In: A systems theoretic approach to systems and synthetic biology I: models and system characterizations. Springer, The Netherlands, pp 307–325
20. Simeonidis E, Chandrasekaran S, Price ND (2013) A guide to integrating transcriptional regulatory and metabolic networks using PROM (probabilistic regulation of metabolism). Methods Mol Biol 985:103–112
21. Chandrasekaran S, Price ND (2013) Metabolic constraint-based refinement of transcriptional regulatory networks. PLoS Comput Biol 9 (12):e1003370
22. Kumar VS, Maranas CD (2009) GrowMatch: an automated method for reconciling in silico/in vivo growth predictions. PLoS Comput Biol 5(3):e1000308. https://doi.org/10.1371/journal.pcbi.1000308
23. Thiele I, Palsson BØ (2010) A protocol for generating a high-quality genome-scale metabolic reconstruction. Nat Protoc 5(1):93–121
24. Feist AM, Herrgård MJ, Thiele I, Reed JL, Palsson BØ (2009) Reconstruction of biochemical networks in microorganisms. Nat Rev Microbiol 7(2):129–143
25. Blais EM, Chavali AK, Papin JA (2013) Linking genome-scale metabolic modeling and genome annotation. Methods Mol Biol 985:61–83
26. Mahadevan R, Schilling CH (2003) The effects of alternate optimal solutions in constraint-based genome-scale metabolic models. Metab Eng 5(4):264–276 doi:S1096717603000582
27. Bonneau R (2008) Learning biological networks: from modules to dynamics. Nat Chem Biol 4(11):658–664. https://doi.org/10.1038/nchembio.122
28. Faith JJ, Hayete B, Thaden JT, Mogno I, Wierzbowski J, Cottarel G, Kasif S, Collins JJ, Gardner TS (2007) Large-scale mapping and validation of Escherichia coli transcriptional regulation from a compendium of expression profiles. PLoS Biol 5(1):e8. https://doi.org/10.1371/journal.pbio.0050008 06-PLBI-RA-0740R3
29. Margolin AA, Nemenman I, Basso K, Wiggins C, Stolovitzky G, Dalla Favera R, Califano A (2006) ARACNE: an algorithm for the reconstruction of gene regulatory networks in a mammalian cellular context. BMC Bioinformatics 7(Suppl 1):S7. https://doi.org/10.1186/1471-2105-7-S1-S7 1471-2105-7-S1-S7
30. Chandrasekaran S, Ament SA, Eddy JA, Rodriguez-Zas SL, Schatz BR, Price ND, Robinson GE (2011) Behavior-specific changes in transcriptional modules lead to distinct and predictable neurogenomic states. Proc Natl Acad Sci U S A 108 (44):18020–18025
31. Karr JR, Sanghvi JC, Macklin DN, Gutschow MV, Jacobs JM, Bolival B Jr, Assad-Garcia N, Glass JI, Covert MW (2012) A whole-cell computational model predicts phenotype from genotype. Cell 150(2):389–401. https://doi.org/10.1016/j.cell.2012.05.044

Chapter 15

^{13}C-Fingerprinting and Metabolic Flux Analysis of Bacterial Metabolisms

Whitney Hollinshead, Lian He, and Yinjie J. Tang

Abstract

^{13}C-assisted metabolism analysis provides rigorous calculations of the intracellular reaction rates (i.e., fluxes) within the central metabolism of microbial hosts. This mapping of the intracellular network within microbes has proven to be essential for understanding the cell physiology. The approach is also a key to identifying central metabolic nodes, probing the rigidity of a metabolic network, revealing cofactor balances, and delineating hidden pathways. Here we present the methodology of using stable isotopic carbon substrates for both qualitative (^{13}C-fingerprinting of functional pathways) and quantitative (Metabolic Flux Analysis) metabolism studies on bacterial species. In this methodology, we include step-by-step instructions to use the open source WUflux software for the steady-state flux calculations based on labeling information of amino acids or free metabolites.

Key words ^{13}C-MFA, Metabolites, Flux ratio, Metabolism, Proteinogenic amino acids, WUflux

1 Introduction

There has been a long history of utilizing isotopic tracers to resolve important questions on cell physiology from DNA synthesis [1] to metabolite synthesis routes. ^{13}C-isotopic tracers have been an essential tool in determining metabolic fluxes, in vivo reaction rates, within the cell central metabolism. This ^{13}C-metabolism analysis aims to decipher the topology and functionality of the cell metabolic network. The insights gained can link the internal network of the microbe with its resulting phenotype. In addition, this fluxomics approach can be integrated with other "omics" approaches such as genomics and transcriptomics to offer a bottom-up approach that exposes internal regulations within the cell under genetic or environmental stresses [2]. The ^{13}C-metabolism analysis can be used in an either qualitative (^{13}C-Fingerprinting) or quantitative (metabolic flux analysis) manner (*see* Fig. 1). ^{13}C-Metabolic flux analysis (^{13}C-MFA) requires measurement of

Christine Nicole S. Santos and Parayil Kumaran Ajikumar (eds.), *Microbial Metabolic Engineering: Methods and Protocols*, Methods in Molecular Biology, vol. 1927, https://doi.org/10.1007/978-1-4939-9142-6_15,

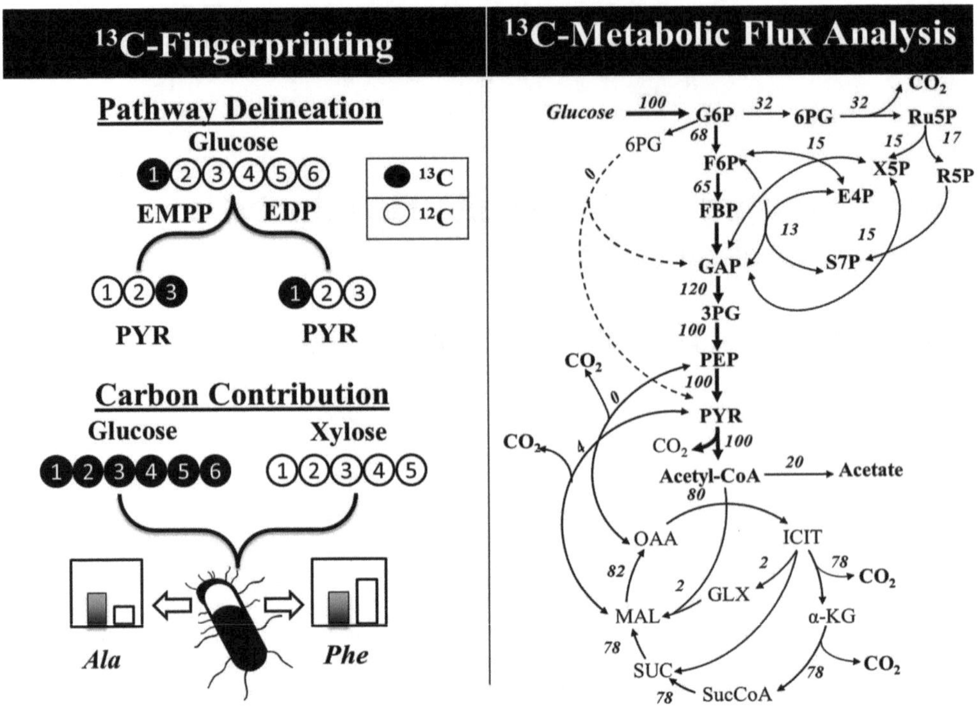

Fig. 1 Qualitative (^{13}C-Fingerprinting) vs. Quantitative (^{13}C-MFA) Analysis. ^{13}C-Fingerprinting investigates functional pathways without rigorous flux calculations, solely relying on simple analysis of the isotopomer data from targeted ^{13}C-isotopic labeling experiments. ^{13}C-MFA resolves the fluxes in the central carbon metabolism using nonlinear least squares regression modeling based on stoichiometric and isotopomer balances

both isotopomer distributions in metabolites and culture growth kinetics. In addition, MFA needs a quality flux model to calculate flux values and the confidence intervals. Therefore, MFA is costly and time consuming. To facilitate characterization of nonmodel species or engineered microbes, ^{13}C-Fingerprinting is a quicker and simpler method that can roughly investigate pathway activities, identify multiple carbon nutrients' contribution to biomass and product synthesis, and calculate true product yields [3, 4].

^{13}C-MFA is the quantitative method that provides precise flux estimates for the reactions within the central carbon metabolism through mathematical modeling. The model assumes cells reach a quasi-steady state (metabolic and isotopic), which limits this method to chemostat cultures or batch cultures during the exponential phase. Using the stoichiometry of the metabolic network, the mass isotopomer distributions (MID) of the metabolites, and the extracellular rates, several available flux software platforms can be used to compute flux distribution in the central metabolism. We will present the protocol of using the WUflux software to compute the fluxes. Despite its wealth of information, traditional ^{13}C-MFA may not be suitable, if the microbial strain fails to reach metabolic steady state, has a poorly annotated metabolic network, or cannot

Table 1
Commonly used tracers for ^{13}C-fingerprinting of pathways [3]

Substrates	Commonly used tracers	Pathway investigations
Glucose	Positional labeled glucose ([1-$^{13}C_1$] Glucose)	EMPP, EDP, and OPPP
Acetate	[1-$^{13}C_1$] Acetate, or [2-$^{13}C_1$] Acetate	TCA cycle and gluconeogenesis
Pyruvate	[U-$^{13}C_3$] Pyruvate, [1-$^{13}C_3$] Pyruvate, [1,2-$^{13}C_2$] Pyruvate, etc.	TCA cycle and gluconeogenesis
Amino acids	[U-$^{13}C_3$] Aspartate, [U-$^{13}C_3$] Glutamate, etc.	Amino acid uptake
Fully labeled substrates	e.g., [U-$^{13}C_2$] Acetate, [U-$^{13}C_6$] Glucose, [U-$^{13}C_6$] lactate	Substrate coutilization
Others	^{13}C-Formate, $^{13}CO_2$ or $NaH^{13}CO_3$	C1 metabolism and CO_2 fixation

be cultured with defined nutrients. In addition, it takes weeks or even months for ^{13}C-MFA, thus it is labor-intensive as compared to the qualitative approach ^{13}C-Fingerprinting. Here, ^{13}C-fingrprinting protocol removes the need for extensive flux calculations by a smart design of ^{13}C-labeling experiments based on the desired goals of the study (*see* Table 1).

2 Materials

2.1 ^{13}C-Labeling Growth Experiment

1. Tracer selections: for probing pathway activities via ^{13}C-fingerprinting or performing metabolic flux analysis, specific ^{13}C-substrates can be selected to achieve the best resolutions. For example, [1-^{13}C] glucose has been extensively used to delineate glycolytic preference among oxidative pentose phosphate pathway (OPPP), Embden–Meyerhof–Parnas Pathway (EMPP), and Entner–Doudoroff pathway (EDP), as these pathways produce distinct labeled pyruvates from glucose [3]. Moreover, fully labeled ^{13}C substrates can be used to examine intrinsic biosynthesis yields of labeled substrates in a medium with complex carbon sources [5]. In Table 1, we list tracers for metabolism analysis of microbial pathways, which are available from chemical manufacturers such as Sigma-Aldrich (St. Louis, MO) or Cambridge Isotopes (Tewksbury, MA).

   For ^{13}C-MFA approaches, using one labeled glucose substrate over a mixture of labeled substrates leads to easier medium preparations and less isotopic noises, with the optimal tracers being [1,6-^{13}C] glucose, [5,6-^{13}C] glucose, or

$[1,2\text{-}^{13}C]$ glucose [6]. Considering the labeling substrate costs, $[1,2\text{-}^{13}C]$ glucose is the recommended choice for ^{13}C-MFA experiments.

2. Culture media: for qualitative metabolic investigations via ^{13}C-fingerprinting, complex media can be used. Rich media often contains complex or even undefined carbon sources such as yeast extract or casamino acids, which will interfere quantitative analysis. Therefore, it is necessary to use a minimal media without any major carbon sources beside the ^{13}C tracer or at least a well-defined media during ^{13}C-MFA. Although it should be avoided, the use of trace vitamin for growth typically has a negligible effect on the results.

2.2 Measurement of Growth and Extracellular Rates

1. UV-Vis Spectrometer.
2. HPLC or enzymatic assay kits for measurement of extracellular consumption and secretion.

2.3 Free Metabolite Extraction

1. Liquid nitrogen.
2. 100 mL Deionized Water.
3. Methanol–chloroform (volume ratio 7:3) solution.
4. Lyophilizer.

2.4 Proteinogenic Amino Acid Extraction

1. Saline solution: 0.9% (wt/vol) NaCl.
2. 6N HCl
3. Oven (capable of reaching 100 °C).
4. Evaporator: Mini-Vap Evaporator/Concentrator (Sigma-Aldrich #22971) or Reacti-Vap Evaporators (ThermoFisher Scientific TS-18825).
5. Drying gas: N_2 or air (*see* **Note 1**).

2.5 Derivatization and GC-MS Analysis

(Free Metabolite Derivatization)

1. MOX reagent: 20 mg/mL methoxyamine hydrochloride in pyridine (prepared fresh).
2. *N*-Methyl-*N*-(trimethylsilyl) trifluoroacetamide (MSTFA) + 1% trimethylchlorosilane (TMCS).
3. GC vials.

or

(Proteinogenic Amino Acid Derivatization)

1. Tetrahydrofuran.
2. *N*-tert-butyldimethylsilyl-*N*-methyltrifluoroacetamide (TBDMS), 97%.

3. GC vials.
4. Gas chromatography (Column HP-5MS or DB-5S)-Mass Spectrometer (Alternative analytical equipment can be used such as NMR; however, be careful to match the instrument with the applicability of your ^{13}C-MFA software.)
5. Amino Acid Standards (Optional) (*see* **Note 2**).

2.6 Flux Calculation Software

Currently, there have been multiple software programs published for correcting MID data and/or flux analysis, for example, 13CFLUX [7], 13CFLUX2 [8], FiatFlux [9], iMS2Flux [10], INCA [11], METRAN [12], OpenFLUX [13], OpenMebius [14], and WUFlux [15]. All the software programs can resolve flux analysis under metabolic and isotopic steady-state conditions. Some of the software programs are open-source, such as OpenFLUX, OpenMebius, and WUFlux, which could be further programmed by users. Some of them can perform isotopically nonstationary flux analysis, such as INCA and OpenMebius, which are useful for analyzing the flux phenotype of autotrophic metabolism. Researchers may choose a software based on the purposes of their studies. Our laboratory has built a ^{13}C-MFA software, WUFlux, featuring different templates for bacterial metabolism. WUFlux has predefined central metabolic network, carbon transition rules of biochemical reactions, straightforward interface and functions for flux map visualization. These features allow for minimum work required from users, and thus make WUFlux straightforward to use for studying bacterial metabolism of common substrates (such as glucose, acetate, and lactate).

3 Methods

^{13}C-based Metabolism analysis is summarized in Fig. 2.

3.1 Knowledge of a Metabolic Network

Metabolism studies start from genome annotations. For example, the KEGG database is a superior resource to reveal metabolic network of interests, which allows the development of quality metabolic models. In addition, the carbon transition of pathways needs to be known as well. This is commonly reserved among organisms and preset in ^{13}C-MFA software. In bacterial central metabolic network, it usually includes glycolysis, pentose phosphate pathway, the TCA cycle, the Entner–Doudoroff pathway, and anaplerotic reactions.

3.2 ^{13}C-Labeling Experiments

1. Prepare a seed culture: Grow the strain in its unlabeled rich medium into the mid-exponential phase.
2. Inoculate into the labeled complex or defined (depending on the application) medium at a low inoculation ratio (<0.5 v/v %) (*see* **Note 3**).

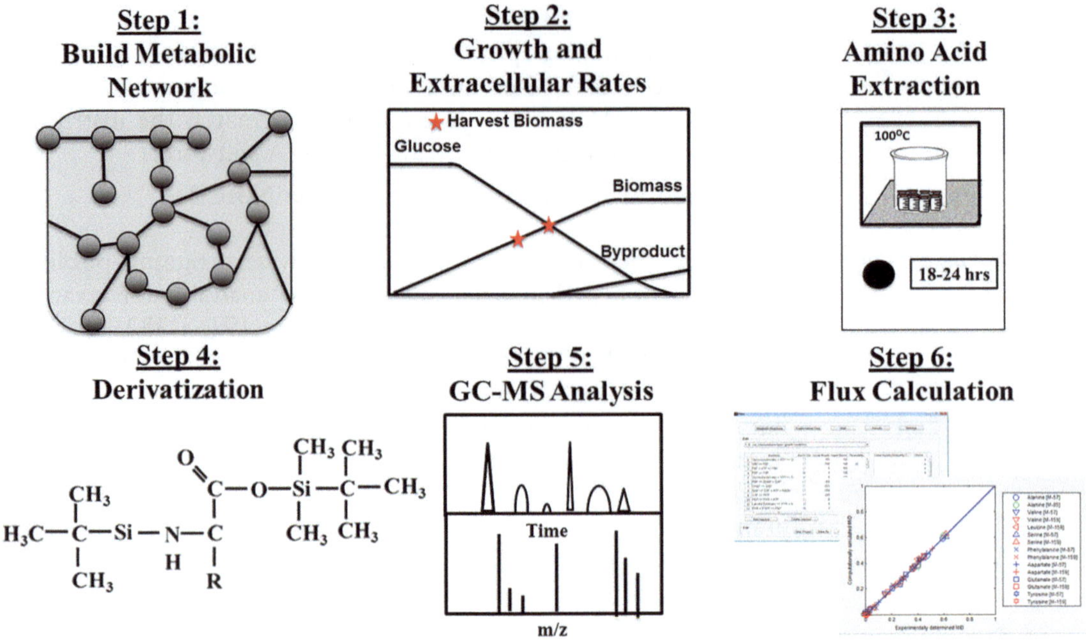

Fig. 2 ^{13}C-metabolism analysis workflow

3. Grow the cells into the mid-exponential phase, and take biomass samples at three or more time points. Place the biomass samples into freezing cold saline solution (1:1) in a centrifuge tube to quench the cell metabolism. The culture can be further cooled to ~0 °C by using a liquid nitrogen bath (*see* **Notes 4** and **5**).
4. Centrifuge the samples (5000 × *g*, 3 min, 0 °C) and keep cell pellets for free metabolite/proteinogenic amino acid extractions. Keep supernatant to determine supernatant glucose or other extracellular metabolite concentrations via enzyme kit or HPLC.

3.3 Free Metabolite Extraction and Analysis

1. Add 1 mL of methanol–chloroform extraction solution to the cell pellet. Important: Keep samples cool throughout the extraction.
2. Shake samples (in GC vials) for 4 h or overnight at 4 °C and 200 rpm.
3. Add 0.5 mL of ultrapure water and vortex samples for 2 min (*see* **Note 6**).
4. Extract the aqueous layer into new GC vials and freeze-dry the samples using a lyophilizer. After this step, the dried samples are stable and can be stored at −20 °C for several days to weeks.
5. Use the LC-MS to analyze free metabolites [17]. If LC-MS is not available, GC-MS may be employed. To derivatize the free

metabolites for GC separations, add 120 μL of MOX reagent and incubate samples at 30 °C for 90 min; then add 180 μL of MSTFA reagent and incubate for 60 min. After fully derivatization, samples should be run on the GC-MS immediately. In general, GC-MS can measure a few free amino acids and organic acids (e.g., succinate and citrate).

3.4 Proteinogenic Amino Acid Extraction and Derivatization [3]

1. Wash cells with 1 mL of saline solution.
2. Add 1.5 mL of 6N HCl to cells, and then transfer into GC-vials. Tightly cap the GC vials, and then place samples in an oven at 100 °C for 18–24 h.
3. Remove hydrolyzed samples from the oven and then let them cool to room temperature. Transfer the samples to microcentrifuge and spin down to remove any cell debris. Evaporate the HCl using an evaporator (*see* **Note 7**). At this step, the dried samples are stable and can be stored at −20 °C for several months.
4. Derivatize the amino acids by adding THF and TBDMS (volume ratio 1:1, and we recommend 150 μL for either solution). Heat in an oven at 70 °C for 1 h. Spin down to remove any debris (*see* **Note 8**).
5. Dilute the sample with additional THF if the samples appear too concentrated. Derivatized samples are relatively stable at 4 °C for 1–3 days, but for best results run samples immediately after derivatization (*see* **Note 9**).

3.5 GC-MS Analysis of Amino Acids

Common GC/MS settings for proteinogenic amino acid are provided in Table 2. Table 3 shows proteinogenic amino acids that can be analyzed by GC/MS. MSTFA-derivatized free metabolites recovered on GC-MS can vary depending on equipment and experimental design, it is best to identify and optimize your GC-MS method using standards.

Table 2
GC/MS settings

Parameters	Settings (amino acid)
Split ratio	1:5 or 1:10
Injection volume	1 μL
Carrier gas	helium = 1.2 mL/min
Program:	150 °C for 2 min 3 °C/min to 280 °C 20 °C/min to 300 °C Hold for 5 min
m/z range:	60–500

Table 3
TBDMS-derivatized amino acids (*see* Note 10)

Amino acid	Precursor	# of carbons	M-159/M-85	M-57
Alanine	PYR	3	158	260
Glycine	3PG, Serine, CO_2 + methylene-THF	2	218 (M-85)	246
Valine	PYR	5	186	288
Leucine	PYR, AcCoA	6	200	N/A
Isoleucine	PYR, OAA	6	200	N/A
Proline	AKG	5	184	286
Methionine	OAA	4	218	320
Serine	3PG, Glycine + methylene-THF	3	288	390
Threonine	OAA, GLX + AcCoA	4	376 (M-85)	404
Phenylalanine	E4P, PEP	9	234	336
Aspartate	OAA	4	316	418
Glutamate	AKG	5	330	432
Lysine	OAA, PYR	6	329	431
Histidine	R5P, Formyl-THF	6	338	440
Tyrosine	E4P, PEP	9	364	466

NA: Overlap peaks. May use M-15 peak to replace M-57

3.6 Correction for Natural Isotope Abundance

The raw mass isotopomer distribution data will have labeling contributions from the natural isotope abundances of the elements (carbon, silicon, etc.) that comprised the derivatized metabolites. The isotopomer data can be corrected for this natural isotope abundance using a software provided through the Wiechert group [16].

3.7 Using ^{13}C-Fingerprinting and Isotopomer Balances to Estimate Pathway Functions

^{13}C-fingerprinting roughly estimates fluxes using simple isotopomer balances. For example, pyruvate (or alanine) labeling, produced through the three different glycolytic pathways from [1-$^{13}C_1$] glucose, can be determined by Eqs. 1–3 [17]:

$$\mathrm{Pyr}_{100} = \frac{v_{\mathrm{EDP}}}{2v_{\mathrm{EDP}} + 2v_{\mathrm{EMP}} + \frac{5}{3}v_{\mathrm{OPPP}}} \tag{1}$$

$$\mathrm{Pyr}_{000} = \frac{\frac{5}{3}v_{\mathrm{OPPP}} + v_{\mathrm{EDP}} + v_{\mathrm{EMPP}}}{2v_{\mathrm{EDP}} + 2v_{\mathrm{EMP}} + \frac{5}{3}v_{\mathrm{OPPP}}} \tag{2}$$

$$\mathrm{Pyr}_{001} = \frac{v_{\mathrm{EMPP}}}{2v_{\mathrm{EDP}} + 2v_{\mathrm{EMPP}} + \frac{5}{3}v_{\mathrm{OPPP}}} \tag{3}$$

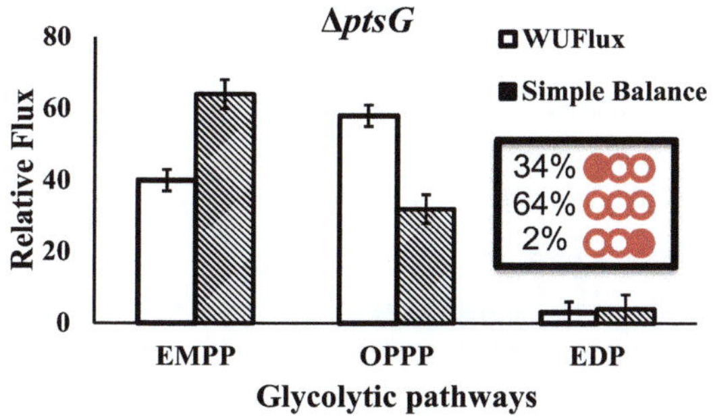

Fig. 3 Comparison of fluxes determined by Eqs. 1–3 and those by ^{13}C MFA. The isotopomer data was generated from ^{13}C-labeling experiment with Δ*ptsG E. coli* mutant using [1-$^{13}C_1$] glucose (Unpublished data)

where v_{EDP}, v_{EMPP}, and v_{OPPP} are the relative fluxes for the ED pathway (EDP), the EMP pathway (EMPP), and the pentose phosphate pathway (OPPP), respectively. I_{xxx} is the fraction of the isotopomer, with 1 denoting the carbon position of ^{13}C (e.g., I_{001} is the abundance of pyruvate labeled at carbon position 3). The above equations can give the relative flux ratios among the EDP, the EMPP, and the OPPP. However, these equations assume no reversibility of reactions and ignore the influence of intersecting pathways (e.g., cataplerotic pathways and biomass formation), thus generating only an approximate analysis of the three glucose catabolic pathways toward pyruvate synthesis. As shown in the graph below (Fig. 3), the flux ratios generated for a Δ*ptsG E. coli* mutant through these simple balances can differ from those calculated by ^{13}C-MFA.

^{13}C-fingerprinting with different labeled tracers is a convenient way to quickly estimate flux ratios or pathway activities without rigorous profiling of global fluxomes. In our previous paper [3], we designed parallel labeling experiments to rapidly investigate microbial pathways through glycolysis, TCA cycle, glyoxylate shunt, and phenolic degradation pathways. In addition, carbon and nutrient contributions to biomass synthesis in mixed carbon media can be revealed with ^{13}C-labeling experiments using a fully labeled primary carbon substrate (e.g., (U-$^{13}C_6$) glucose) with unlabeled secondary substrates (e.g., yeast extract). The resulting ^{13}C enrichment into biomass would represent the contribution (or intrinsic yield) from the labeled carbon source.

3.8 Flux Calculations by Using WUFlux

WUflux is an open source MATLAB based flux calculation platform developed by Dr. Lian He from the Tang Lab. Here, we demonstrate the general procedures of performing ^{13}C MFA with WUFlux. For the detailed information, please read the user manual of WUFlux at http://www.13cmfa.org/.

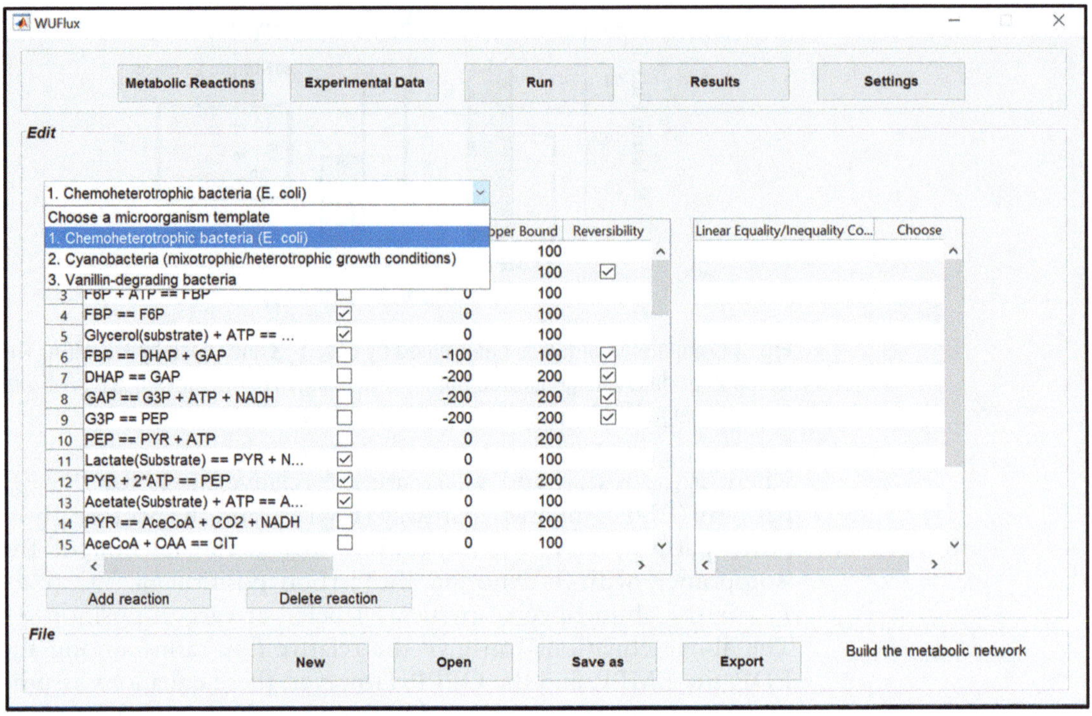

Fig. 4 Template selection in WUFlux

1. *Choose a template.* Click "Metabolic Reactions," and then select one template from those for chemoheterotrophic bacteria, photomixotrophic/photoheterotrophic cyanobacteria, or vanillin-degrading bacteria. Figure 4 shows a screenshot of choosing a model template in WUFlux.
2. *Edit metabolic reactions.* WUFlux displays central metabolic reactions, which includes the EMPP, the OPPP, the EDP, the TCA cycle, the Calvin cycle (only in the cyanobacterial template), the C1 (including 5-methyl-tetrahydrofolate and 5,10-methylene-tetrahydrofolate) metabolism pathway and the anaplerotic pathway. Researchers can modify the metabolic network by knocking out reactions that are not present in the bacterial metabolism, changing the boundary conditions and defining the reversibility of reactions based on the gene annotations and reaction thermodynamics (*see* **Note 11**).
3. *Input experimental data.* To import experimental data, click the button "Experimental Data." Two types of experimental data are required: the MIDs of the metabolites and the labeled substrates used. For input of the MIDs, users can either import the data from an Excel file, or input the data manually. For input of labeled carbon substrates, a list of common carbon substrates is provided under labeled tracers. For your chosen

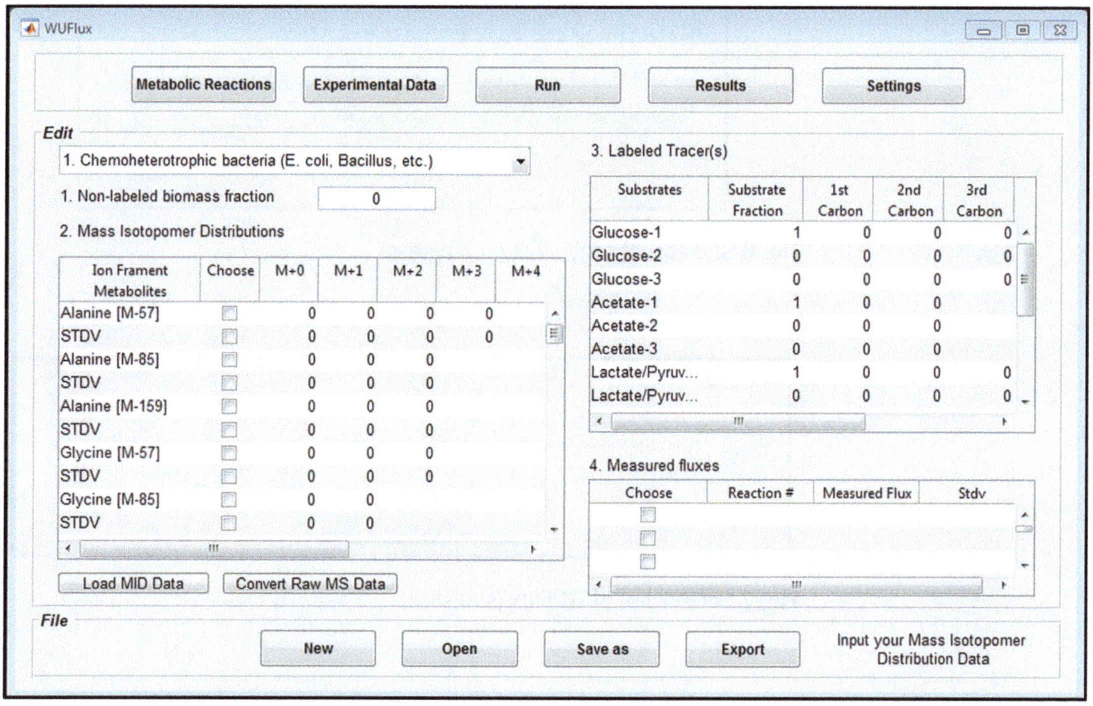

Fig. 5 'Experimental Data' in WUFlux

carbon substrate, each carbon atom of labeled substrate should be defined as the relative abundance of ^{13}C atom. For example, define [1,2-$^{13}C_2$] glucose as [1,1,0,0,0,0] for the corresponding six carbon atoms, where the first "1" means the first carbon atom of glucose is 100% ^{13}C-carbon. In addition, two optional data types can also be included: the nonlabeled biomass percentage and the measured fluxes. The former is used to remove the noise from nonlabeled ^{12}C-carbon introduced from the bacterial inoculation, and the latter can help accurately determine the intracellular fluxes. We recommend users to consider these two factors when the initial nonlabeled biomass in the labeled culture is high or any by-product is significantly produced by the microorganism. Figure 5 shows the experimental inputs in WUFlux.

4. *Customize optimization settings*. To modify parameters for optimization and statistics, click "Settings" and then make changes. The MATLAB built-in function "fmincon" is used to solve the nonlinear optimization problem. The parameters, TolFun, TolCon, TolX, MaxIter, and MaxFun, have the same meanings as described on the website: http://www.mathworks.com/help/optim/ug/fmincon.html. The default algorithm of the "fmincon" solver is "interior-point." To get the global optimum, ten or multiple initial guesses are recommended. In

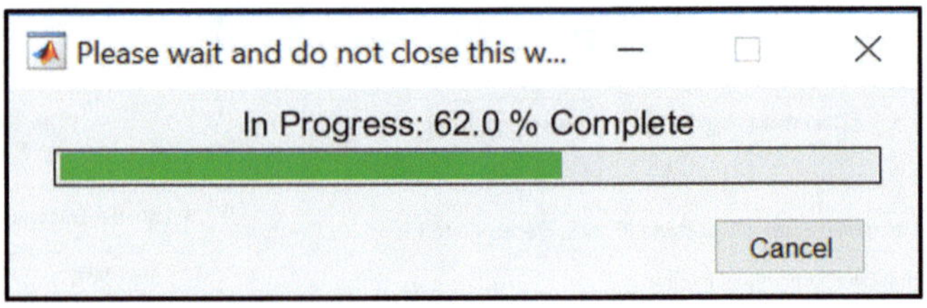

Fig. 6 Screenshot of WUFlux (in progress)

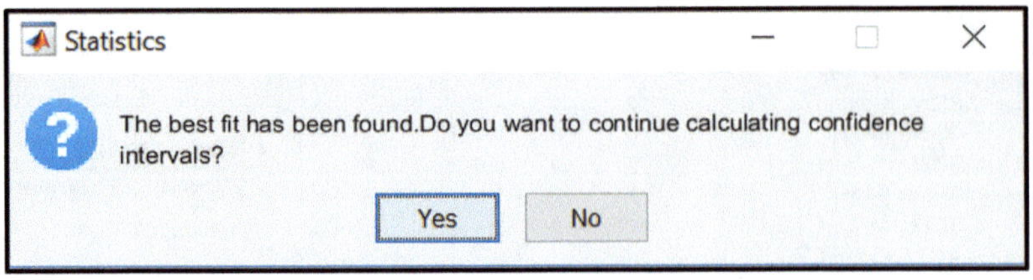

Fig. 7 Screenshot of WUFlux (statistics)

addition, the Monte Carlo method can calculate the confidence interval of fluxes. In this method, WUFlux perturbs the measured data with normally distributed noises, whose standard deviations (δ) are based on the experimental results.

5. *Optimize the flux distributions.* To calculate the intracellular flux distributions, click "Run," and the following pop-up window (Fig. 6) will appear to show the progress of the ongoing calculation. All the flux calculations starting from each initial guess will be stored in the file.

   Once finished, another window (Fig. 7) will appear:
   Either click "Yes" to continue calculating confidence intervals, or click "No" to stop the calculation for the time being. The "Start" button changes to the "Continue" button, and you can resume calculating confidence intervals later.

6. *Visualize the results.* Click "Results" to see the best fit of flux distributions, confidence intervals, and standard deviations. Figure 8 showing the correlation of measured and simulated MID data will be automatically generated on the right side. Additionally, the χ^2 test can be applied to determine the goodness of fit (shown at the bottom line), which can be used as a reference to determine whether the final fitting is statistically acceptable. The interface also has a "Remove results" button, which allows you to remove the current flux results. However, once deleted, the results cannot be recovered.

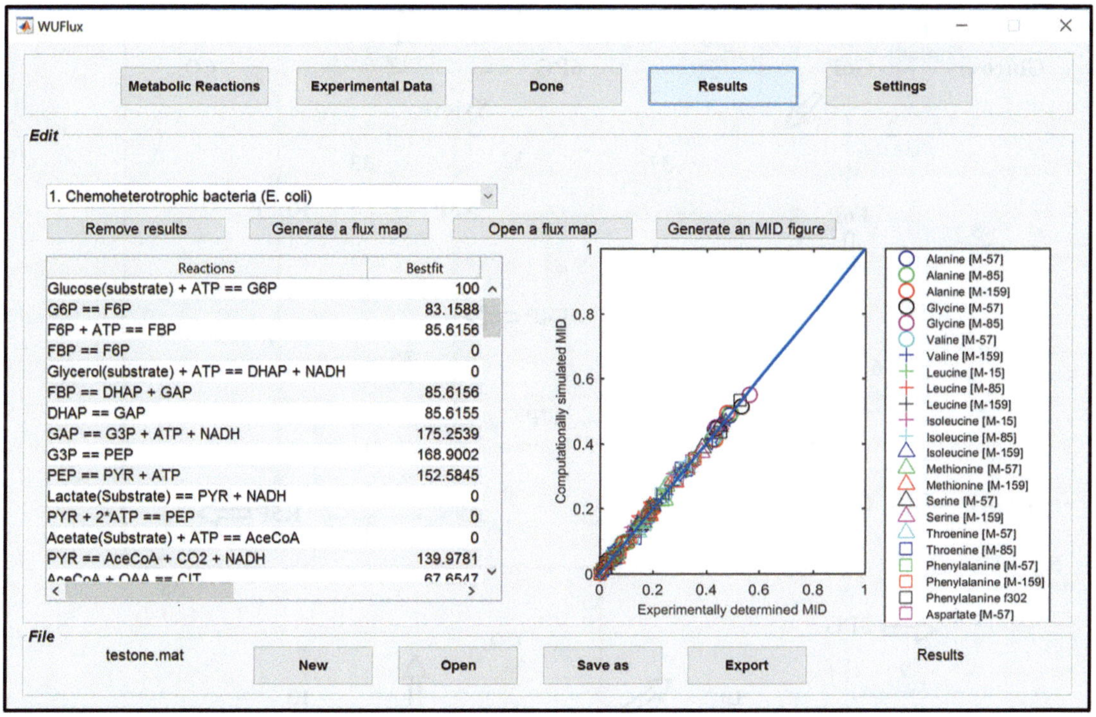

Fig. 8 "Results" in WUFlux

To visualize the flux results with respect to a constructed metabolic network, click the button "Generate a flux map." WUFlux will generate a flux map (Fig. 9) in a .svg file. Further, to see the comparison between the simulated and experimental MID of a certain ion fragment, click "Generate an MID figure" and then choose the corresponding ion fragment.

4 Notes

1. N_2, an inert gas, is the preferred drying gas, though air is a cheaper alternative.
2. Standards can be helpful for accurate peak identification.
3. For qualitative analysis, the inoculation ratio can be higher than 1% (v/v).
4. Typically for *E. coli*, we harvest 5 mL of cultures at OD_{600} ~1.0. Only 5 mg of biomass is necessary for amino acid detection via GC-MS analysis. Biomass can be stored at −20 °C.
5. Due to the fast turnover of free metabolites, it is best to quench the cells (i.e., halt the metabolism), especially for fast turnover metabolites (not necessary for proteinogenic amino acids). Quenching can be completed through fast filtration, cold

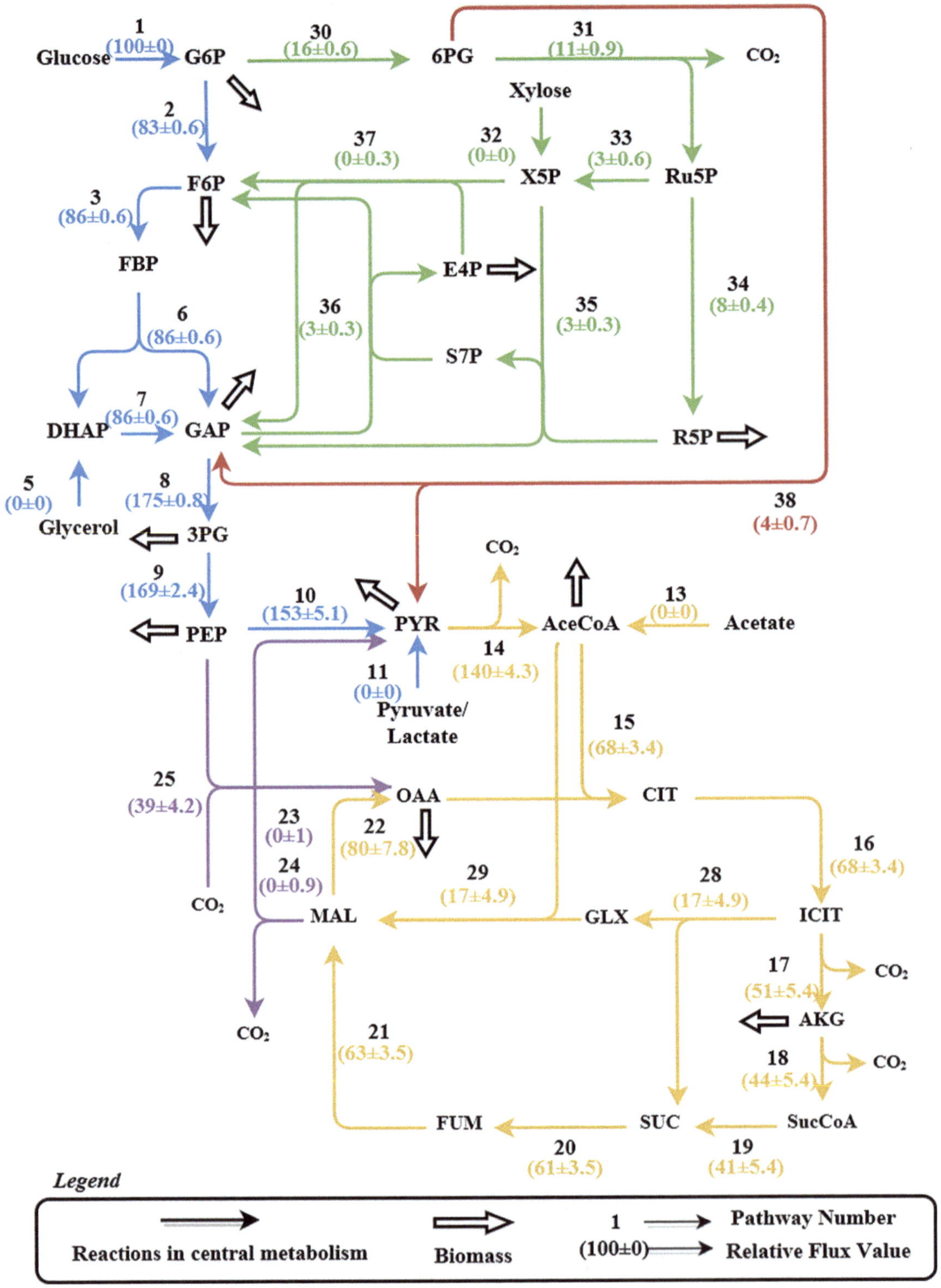

Fig. 9 A representative flux distribution map generated by WUFlux

solvent (ethanol or methanol), or cold media. Our preferred method is a combination of cold media (no carbon sources) with liquid nitrogen bath as a coolant.

6. Samples should form an aqueous and an organic layer.
7. If the samples appear a dark brown/black color, it is likely that the biomass amount was too high. At this stage, you can dry less volume of hydrolyzed sample.
8. Large particles will clog the needle of the GC-MS, so closely examine samples prior to GC-MS run for any remaining cell debris.
9. It is ideal to avoid loading concentrated samples; this will cause oversaturated MS signals which leads to an inaccurate isotopomer data. Typically, the derivatized sample should be a yellow/light brown. A dark color denotes an overly concentrated sample, which requires dilution of amino acid samples using THF or decreasing the split ratio of GC injection.
10. Cysteine and tryptophan cannot be measured through this technique. Glutamine and asparagine are converted in this method to glutamate and aspartate, respectively. Due to their low abundance, these amino acids may be difficult to analyze in dilute samples: methionine, threonine, arginine, and histidine. In addition, if peaks appear to coelute with other compounds, it may be necessary to decrease the temperature ramp for better elucidation. General signs of interference from other compounds are a high M-1 peak.
11. In WUFlux, a new reaction in the central metabolism cannot be added in the template, and all the carbon transition fates have been predefined. But users can add product formation reactions in the "Metabolic Reactions" section.

Acknowledgments

We thank the funding support from DOE (DESC0018324).

References

1. Meselson M, Stahl FW (1958) The replication of DNA in Escherichia Coli. Proc Natl Acad Sci U S A 44:671–682
2. He L, Xiao Y, Gebreselassie N, Zhang F, Antoniewicz MR, Tang YJ, Peng L (2014) Central metabolic responses to the overproduction of fatty acids in *Escherichia coli* based on ^{13}C-metabolic flux analysis. Biotechnol Bioeng 111:575–585
3. Hollinshead WD, Henson WR, Abernathy M, Moon TS, Tang YJ (2016) Rapid metabolic analysis of Rhodococcus opacus PD630 via parallel 13C-metabolite fingerprinting. Biotechnol Bioeng 113: 91–100
4. Varman AM, He L, You L, Hollinshead W, Tang YJ (2014) Elucidation of intrinsic biosynthesis yields using 13C-based metabolism analysis. Microb Cell Factories 13:42
5. Ruan Z, Hollinshead W, Isaguirre C, Tang YJ, Liao W, Liu Y (2015) Effects of inhibitory compounds in lignocellulosic hydrolysates on

Mortierella isabellina growth and carbon utilization. Bioresour Technol 183:18–24
6. Crown SB, Long CP, Antoniewicz MR (2016) Optimal tracers for parallel labeling experiments and 13C metabolic flux analysis: a new precision and synergy scoring system. Metab Eng 38:10–18
7. Wiechert W, Mollney M, Petersen S, de Graaf A (2001) A universal framework for ^{13}C metabolic flux analysis. Metab Eng 3:265–283
8. Weitzel M, Noh K, Dalman T, Niedenfuhr S, Stute B, Wiechert W (2013) ^{13}CFLUX2–high-performance software suite for ^{13}C-metabolic flux analysis. Bioinformatics 29:143–145
9. Zamboni N, Fischer E, Sauer U (2005) FiatFlux-a software for metabolic flux analysis from ^{13}C-glucose experiments. BMC Bioinformatics 6:209
10. Poskar CH, Huege J, Krach C, Franke M, Shachar-Hill Y, Junker B (2012) iMS2Flux–a high-throughput processing tool for stable isotope labeled mass spectrometric data used for metabolic flux analysis. BMC Bioinformatics 13:295
11. Young JD (2014) INCA: a computational platform for isotopically non-stationary metabolic flux analysis. Bioinformatics 30:1333–1335
12. Yoo H, Antoniewicz MR, Stephanopoulos G, Kelleher JK (2008) Quantifying reductive carboxylation flux of glutamine to lipid in a brown adipocyte cell line. J Biol Chem 283:20621–20627
13. Quek L-E, Wittmann C, Nielsen LK, Krömer JO (2009) OpenFLUX: efficient modelling software for ^{13}C-based metabolic flux analysis. Microb Cell Factories 8:25
14. Kajihata S, Furusawa C, Matsuda F, Shimizu H (2014) OpenMebius: an open source software for isotopically nonstationary ^{13}C-based metabolic flux analysis. Biomed Res Int 2014:10
15. He L, Wu SG, Zhang M, Chen Y, Tang YJ (2016) WUFlux: an open-source platform for 13C metabolic flux analysis of bacterial metabolism. BMC Bioinformatics 17:444
16. Wahl SA, Dauner M, Wiechert W (2004) New tools for mass isotopomer data evaluation in (13)C flux analysis: mass isotope correction, data consistency checking, and precursor relationships. Biotechnol Bioeng 85:259–268
17. Hollinshead WD, Rodriguez S, Martin HG, Wang G, Baidoo EE, Sale KL, Keasling JD, Mukhopadhyay A, Tang YJ (2016) Examining *Escherichia coli* glycolytic pathways, catabolite repression, and metabolite channeling using *Δ pfk* mutants. Biotechnol Biofuels 9:212

Chapter 16

A Genetic Screen for the Isolation of Mutants with Increased Flux in the Isoprenoid Pathway of Yeast

Manisha Wadhwa and Anand K. Bachhawat

Abstract

The yeast *Saccharomyces cerevisiae* is one of the preferred hosts for the production of terpenoids through metabolic engineering. A genetic screen to identify novel mutants that can increase the flux in the isoprenoid pathway has been lacking. We present here the method that has led to the development of a carotenoid based visual screen by exploiting the carotenogenic genes from the red yeast *Rhodosporidium toruloides,* an organism known to have high levels of carotenoids. We also discuss the methods to use this screen for the identification of mutants that can lead to higher isoprenoid flux. The carotenoid based screen was developed in *S. cerevisiae* using phytoene synthase *Rt*PSY1 and a hyperactive mutant of the enzyme phytoene dehydrogenase, *Rt*CRTI(A393T) from *Rhodosporidium toruloides.* As validation of the genetic screen is critical at all stages, we describe the method to validate the screen using a known flux increasing gene, a truncated HMG1 (*tHMG1*). To demonstrate how this screen can be exploited to isolate mutants, we described how targeted mutagenesis of candidate gene, *SPT15* a TATA binding protein involved in the global transcription machinery can be carried out to yield novel mutants with increased metabolic flux. Since it is also important to ensure that the isolated mutants are enhancing general isoprenoid flux, we describe how this can be established using an alternate isoprenoid, α-farnesene.

Key words *S. cerevisiae*, *R. toruloides*, Phytoene synthase, Phytoene dehydrogenase, *SPT15*, Mutagenesis, Carotenoids

1 Introduction

Isoprenoids or terpenoids constitute one of the largest group of natural products and are built from five-carbon isoprene units [1]. Production of these compounds from microbial hosts such as *E. coli* or *S. cerevisiae*, when feasible, is a superior option to their extraction from the natural sources. The biosynthetic enzymes and the cytochrome P450s often involved in the formation of these isoprenoids in plants are often better expressed in yeasts, making them in many cases, a preferred host. An important goal in metabolic engineering is to increase the yield of these terpenoids during the heterologous production. The mevalonate pathway for

Christine Nicole S. Santos and Parayil Kumaran Ajikumar (eds.), *Microbial Metabolic Engineering: Methods and Protocols*, Methods in Molecular Biology, vol. 1927, https://doi.org/10.1007/978-1-4939-9142-6_16,

isoprenoid biosynthesis in yeast has been manipulated to increase the flux through the pathway, by removing regulatory blocks, and rate-limiting steps and diversion to alternate pathways. However, it is possible that unknown pathways may also have an impact on the yields of these isoprenoids. The identification of such mutants/genes with increased isoprenoid flux is therefore required, but it requires a good screening strategy that is both robust and high throughput [2]. However, as many of the isoprenoid molecules are water insoluble, volatile and toxic to the production host, a screening strategy/method for increasing isoprenoid yields is quite challenging.

Carotenoid color and intensity provides a visual method for fast and easy detection. The use of carotenoids in a variety of genetic screens dates back to 1994 and has been utilized by many groups for isolation of mutant variants of enzymes [3–7] and identification of novel carotenoids [8, 9]. Carotenoid based genetic screens to increase the overall flux in the isoprenoid pathway have also been attempted [10, 11]. In these studies, the genes responsible for carotenoid biosynthesis, namely, geranyl geranyl diphosphate synthase (GGPP synthase), phytoene synthase, and phytoene dehydrogenase, from *Xanthophyllomyces dendrorhous* were heterologously expressed in *S. cerevisiae*. This carotenoid producing yeast strain was then transformed with mutagenized plasmid/knockout library and thus screened for colonies with increased pigmentation. While yielding mutants that increased carotenoids, they failed to yield mutants or genes that increased general isoprenoid flux. It was observed in these cases that when flux was increased in the isoprenoid pathway, through a known flux increaser, such as *tHMG1*, a surprising decrease in pigmentation (rather than increase) was observed [12]. This decrease in color has been explained to be due to the larger accumulation of the colorless intermediate, phytoene.

A critical factor hampering the previous attempts has been the accumulation of the colorless precursor phytoene that masks any increase in colored carotenoids. Keeping this observation in mind we have devised strategies to obtain a desired screen for increasing general isoprenoid flux. For this we have exploited the carotenogenic enzymes from an organism that is known to have high levels of β-carotene thereby enhancing the pulling capacity of flux downstream of phytoene. Secondly, we have isolated catalytically active mutant of the rate limiting enzyme phytoene dehydrogenase by directed evolution. Thirdly, we have deployed a less efficient GGPP synthase, the native *S. cerevisiae* enzyme, *BTS1*, which when expressed led to lower levels of phytoene. Successful development of the screen at all the stages was monitored by *tHMG1* (truncated 3-hydroxy-3-methyl glutaryl CoA reductase) as a known flux enhancer.

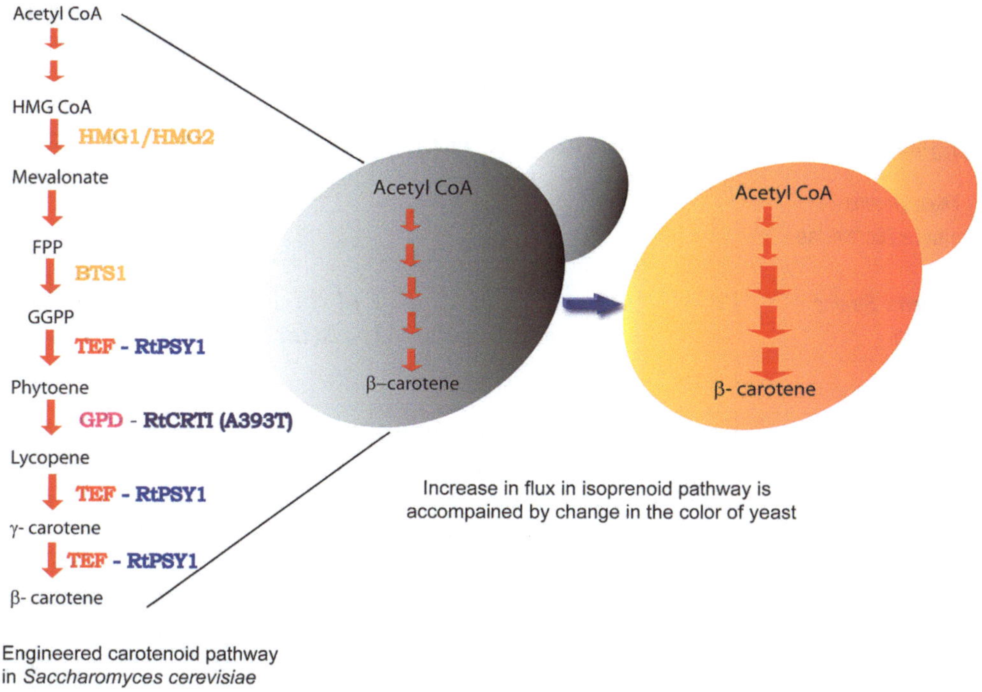

Fig. 1 Schematic representation of developed genetic screen using heterologously engineered carotenoid pathway of *Rhodosporidium toruloides* in *Saccharomyces cerevisiae*

By using above mentioned approaches, we have developed a new carotenoid based genetic screen [13, 14]. The source organism for carotenogenic enzymes was *Rhodosporidium toruloides*. The carotenogenic enzymes in the genetic screen was phytoene synthase (*Rt*PSY1) and mutant phytoene dehydrogenase (*Rt*CRTI (A393T)) and were expressed under strong and constitutive promoters—TEF and GPD respectively. The developed screen was also validated by *tHMG1*. We subjected SPT15, a global TATA binding protein, to directed evolution and isolated isoprenoid flux increasing mutants of *SPT15* using the developed genetic screen. These *SPT15* mutants also enhanced the levels of an alternate isoprenoid, α-farnesene confirming the overall increase in the flux in the isoprenoid pathway. Schematic representation of the developed genetic strain is shown in Fig. 1. This chapter describes the key methods for the development, validation and the exploitation of the screen for identifying new mutants.

2 Materials

Growth media and stock solutions should be prepared in distilled water and autoclaved at 121 °C for 15 min using a liquid cycle (*see* **Note 1**). After autoclaving the material, store the liquid media and solutions at room temperature (exceptions are specified in the text).

All solutions should be prepared using analytical grade reagents and all waste disposal regulations should be followed while discarding the toxic waste materials.

2.1 Strains

2.1.1 Yeast Strains—Saccharomyces cerevisiae

S. cerevisiae ABC 276: *MATα ura3-52 leu2Δ1 his3Δ200 trp1 lys2-80* (a derivative of S288c).

2.1.2 Bacterial Strains—Escherichia coli

E. coli DH5α: F^- *endA1 glnV44 thi-1 recA1 relA1 gyrA96 deoR nupG purB20* φ80d *lacZ*ΔM15 Δ(*lacZYA-argF*) U169, hsdR17 ($r_K^- m_K^+$), λ^-.

2.2 Growth Media

2.2.1 YPD (Yeast Extract, Peptone, Dextrose–Glucose)

YPD is composed of 1% (w/v) Bacto yeast extract, 2% (w/v) Bacto peptone, and 2% (w/v) glucose. For preparation of 500 mL YPD, weigh 5 g Bacto yeast extract, 10 g Bacto peptone, and 10 g glucose and transfer it to 1 L-flask containing 300 mL distilled water and stir the solution on magnetic stirrer. After mixing components, make up the volume of solution to 500 mL with distilled water. YPD agar plates are also prepared in the same way but after stirring and making up the volume of the solution, add 2% agar (w/v). After autoclaving, cool the solution to about 60 °C and pour approximately 30 mL of the YPD-agar media to sterile 90 mm petri plate in a laminar hood and let it cool/solidify. Store the plates at 4 °C.

2.2.2 LB (Luria–Bertani) and LB-amp Plates

LB is composed of 0.5% (w/v) Bacto yeast extract, 1% (w/v) tryptone, and 1% (w/v) NaCl. For preparation of 500 mL LB media, weigh 2.5 g Bacto yeast extract, 5 g tryptone, and 5 g NaCl and transfer it to 1 L flask containing 300 mL distilled water. Stir, and make up the volume to 500 mL with distilled water. LB Amp agar plates are also prepared in the same way but after making up the volume of the solution, add 2% (w/v) agar. After autoclaving, cool the solution to about 60 °C and add 0.5 mL of ampicillin solution from 100 mg/mL stock solution (Final concentration 100 μg/mL) (*see* **Note 2**). Pour approximately 30 mL of the LB-amp agar media to sterile 90 mm petri plate in a laminar hood and let it cool/solidify. Store the plates at 4 °C.

2.2.3 SD+Supplements (Synthetically Defined Supplement Mixture)

This media contains 1.7 g/L yeast nitrogen base (without amino acids and ammonium sulfate) Difco, USA, 5 g/L ammonium sulfate, and 20 g/L glucose plus amino acid supplements as required. For preparation of 500 mL SD media, weigh 0.85 g yeast nitrogen base, 2.5 g of ammonium sulfate, and 10 g glucose and transfer it to 1 L flask containing 300 mL distilled water and add required amino acids at concentration of 80 mg/L and stir the mixture on magnetic stirrer and make up the volume of the solution to 500 mL with distilled water. SD agar plates are also prepared as SD media, but after mixing the solution, add 2.2% (w/v) agar to the solution. After autoclaving, cool the solution to about 60 °C and pour approximately 30 mL of the SD-agar media to sterile 90 mm petri plate in a laminar hood and let it cool/solidify. Store the plates at 4 °C.

2.3 Reagent Solutions

2.3.1 10× TE (pH 7.4)

A 10× TE solution contains 100 mM Tris–HCl (pH 7.4) and 10 mM EDTA.

For preparation of 500 mL 10× TE solution in 1 L flask, transfer 50 mL of 1 M Tris–HCl pH 7.4, 10 mL of 0.5 mM EDTA (pH 8.0) and make up the volume to 500 mL with distilled water. Autoclave the solution and store it at 4 °C after autoclaving.

2.3.2 10× Lithium Acetate (1M)

For preparation of 500 mL Lithium acetate solution, weigh 51.01 g lithium acetate dihydrate (Catalog no. L6883, MW 102.02, Sigma-Aldrich, India) and make up the solution to 500 mL in a 1 L flask with distilled water. Dissolve the solution on the magnetic stirrer and autoclave.

2.3.3 50% (w/v) Poly (Ethylene Glycol)

Weigh 50 g polyethylene glycol (Cat. no. P4338, mol wt. 3350, Sigma-Aldrich, India) and initially add 50 mL 1× TE (pH 7.4) in a 250 mL flask. Stir the solution on magnetic stirrer. After mixing, make up the volume to 100 mL with 1× TE (pH 7.4) and autoclave (*see* **Note 3**).

2.3.4 1M Hydroxylamine Hydrochloride (pH 6.7)

For preparation of 10 mL of solution, weigh 0.35 g hydroxylamine hydrochloride and transfer it to 15 mL falcon tube containing 4.55 mL ice-cold sterile milli Q water. Add 450 μL 5 M NaOH solution. Adjust the pH of solution to 6.7 and keep the solution on ice till use. Prepare the solution fresh immediately/just before use (*see* **Note 4**).

2.4 Kits Used

Gel extraction kit from Thermo Fisher Scientific.

Plasmid miniprep kit from Thermo Fisher Scientific.

2.5 Equipment Required

Laminar air flow.

PCR machine/Thermocycler.

Incubator shaker set at 37 °C for bacterial culture and 30 °C for yeast culture.

Gas chromatography with flame ionization detector (GC-FID).

3 Methods

The development of the screen is first described here and followed by the use of the screen for the isolation of mutants using a candidate gene approach. A schematic representation of the protocol is shown in Fig. 2.

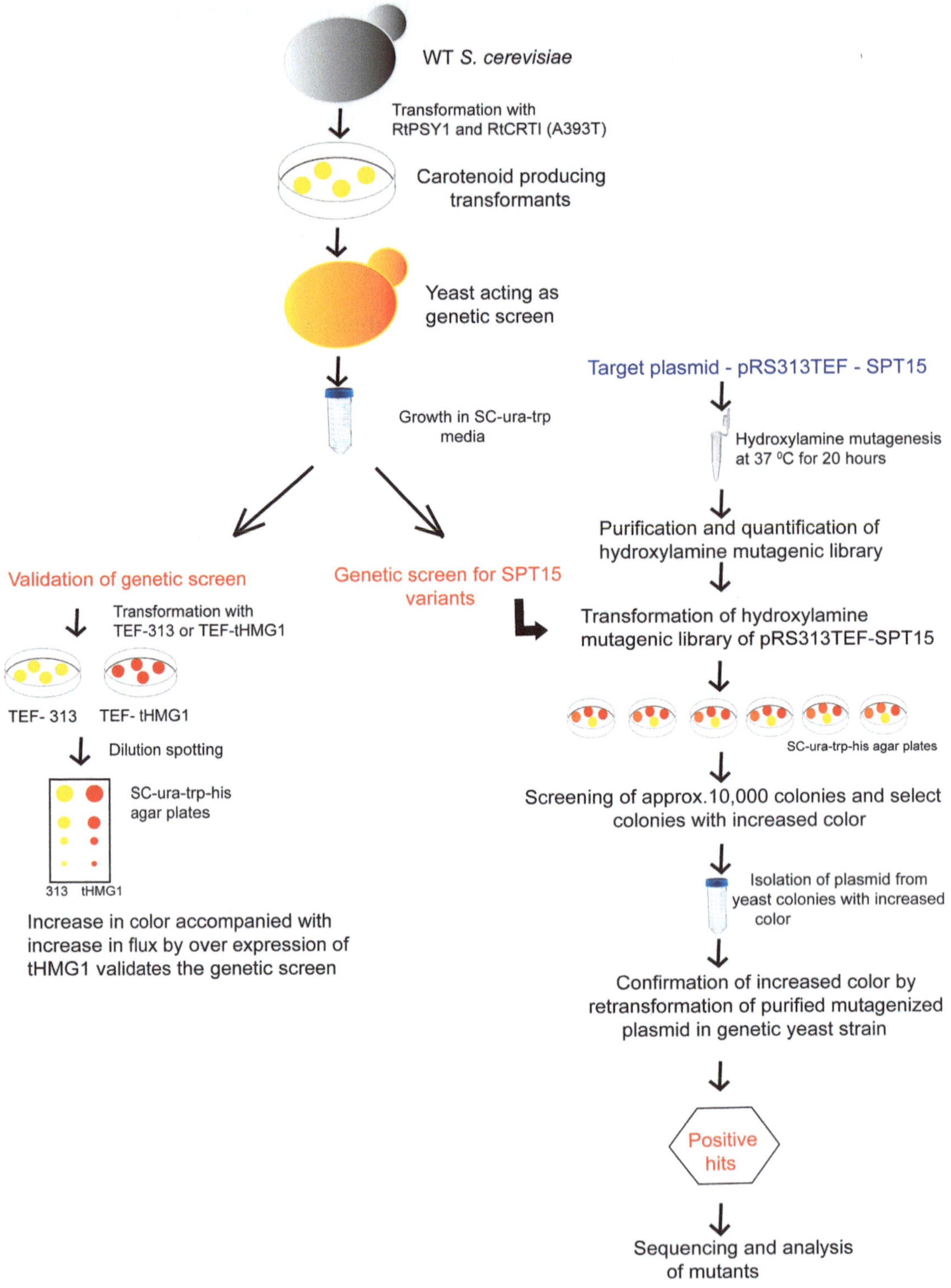

Fig. 2 Schematic representation of protocols for development of a genetic screen and isolation of mutants by directed evolution

3.1 Construction of Carotenoid Producing S. cerevisiae Strain Using Core Carotenogenic Enzymes from R. toruloides

3.1.1 Custom Synthesis and Cloning of Phytoene Synthase and Phytoene Dehydrogenase from Rhodosporidium toruloides

The phytoene synthase gene (*Rt*PSY1) of *R. toruloides* has a length of 2258 bp that includes eight exons and seven introns. The cDNA length is 1839 bp which encodes a protein of 612 aa. The protein has 45% similarity to the phytoene synthase of *Xanthophyllomyces dendrorhous*.

The phytoene dehydrogenase (*Rt*CRTI) of *R. toruloides* has a length of 2577 bp that includes ten exons and nine introns. The cDNA length is 1833 bp which encodes a protein of 610 aa. Sequence alignment of the phytoene dehydrogenase protein from *R. toruloides* with phytoene dehydrogenase from *X. dendrorhous* and other organisms, suggests that N-terminal has extra 56 aa may be resulting from misannotation of start site. Therefore, we consider cDNA of 1665 bp length which encodes protein of 554 aa. It has similarity of 67% to phytoene desaturase (*CRTI*) from *X. dendrorhous*.

Owing to high GC content, the genes were custom synthesized from GenScript, USA after codon optimization using EnCor Biotechnology Inc. (http://www.encorbio.com/protocols/Codon.htm) software.

*Rt*PSY1 cDNA is custom-synthesized with *Bam*HI and *Xho*I restriction site at 5′ and 3′ end respectively. cDNA is digested with *Bam*HI and *Xho*I enzyme and column purified. It is cloned in pRS416TEF vector which is a single copy yeast centromeric vector that contains the strong constitutive TEF promoter and URA3 selection marker. *Rt*PSY1 cDNA is cloned downstream of TEF promoter. *Rt*CRTI cDNA is custom synthesized with *Bam*HI and *Sal*I restriction site at 5′ and 3′ end respectively. cDNA is digested with *Bam*HI and *Sal*I enzyme and column purified. It is cloned in pRS314GPD vector which is a single copy yeast centromeric vector that contains the strong constitutive GPD promoter and tryptophan selection marker. *Rt*CRTI cDNA is cloned downstream of GPD promoter. Ligation of restriction enzyme digested and column purified *Rt*CRTI cDNA and pRS314GPD vector is performed at 16 °C for 16 h (*see* **Note 5**). Ligated product is transformed in competent *E. coli* DH5α cells and plates are kept at 37 °C. Transformants are screened by restriction digestion. Glycerol stock is made in 50% glycerol and stored at −80 °C.

*Rt*CRTI (A393T) mutant of *Rt*CRTI is obtained after directed evolution and subcloned in pRS314GPD. *E. coli* transformant containing pRS314GPD- RtCRTI(A393T) is also stored at −80 °C after making glycerol stock in 50% glycerol.

3.1.2 Transformation of S. cerevisiae with Carotenogenic Genes

1. *E. coli* transformants containing p416TEF- RtPSY1 and pRS314GPD- RtCRTI (A393T) is streaked on LB amp plates and plasmids are isolated from 5 mL of culture using plasmid

miniprep kit and concentration is checked by spectrophotometer by measuring OD_{260} (1 OD_{260} = 50 μg of ds DNA).

2. *S. cerevisiae* ABC 276 strain is streaked onto YPD agar plate from −80 °C stock and plate is kept at 30 °C and after obtaining single colonies on YPD-agar plate, a single colony is inoculated in 5 mL YPD liquid media and incubated at 30 °C on rotary shaker.
3. After the overnight growth, the optical density of culture is measured at 600 nm (OD_{600}). The culture is reinoculated into 10 mL of fresh YPD liquid media to obtain a final $OD_{600} = 0.1$ and incubate at 30 °C with orbital agitation 200 rpm.
4. Incubate the culture till the OD_{600} reaches to approx. 1, centrifuge the cell suspension at 2800 × *g* for 3 min at room temperature.
5. Discard the supernatant and wash the pellet with autoclaved distilled water.
6. Centrifuge at 2800 × *g* for 3 min and wash the pellet with solution containing 1× Lithium acetate made in 1× TE (pH 7.4) (freshly prepared) and centrifuge at 2800 × *g* for 3 min.
7. Resuspend the pellet in 1× lithium acetate–TE solution (prepared in previous step) and incubate at 30 °C with orbital agitation 200 rpm for 30 min.
8. Pellet the culture after 30 min and resuspend the pellet in 200 μL of 1× Lithium acetate–TE solution and add 800–1000 ng of plasmids—p416TEF-*Rt*PSY1 and pRS314GPD-*Rt*CRTI(A393T), 10 μL of ssDNA (single stranded DNA from salmon sperm) from stock solution (10 mg/mL), and 280 μL of 50% PEG. Mix the suspension and kept at 30 °C for 45 min without agitation.
9. After 45 min, heat-shock the cell suspension mixture at 42 °C for 10 min. After that, centrifuge the suspension mixture at 2800 × *g* for 3 min and discard the supernatant and wash the pellet with autoclaved distilled water and resuspend the pellet in 100 μL autoclaved distilled water and spread it on SD plates supplemented with appropriate amino acids but lacking uracil and tryptophan. Incubate the plates at 30 °C.

Table 1
List of primers used in the study. Restriction sites of enzymes are underlined

Primer name	Sequence (5′–3′)
tHMG1 FP	GATCGCGGATCCATGGACCAATTGGTGAAAACTGAAG
tHMG1 RP	CATGCGCCCGGGTTAGGATTTAATGCAGGTGACG
SPT15 FP	GATCGCGGATCCATGGCCGATGAGGAACGTTTAAAG
SPT15 RP	CATGCGCTCGAGTCACATTTTTCTAAATTCAC
*At*FS FP	GACGTTCTAGAATGCCTAAACGACAGGCTCAAC
*At*FS RP	GGCTCGGATCCTTAATTGAGTGGAAGAGGGTGG

3.2 To Evaluate the Increase in Color of Genetic Screen with a Known Flux Increaser, tHMG1

3.2.1 Cloning of tHMG1 and Transformation in Carotenoid Producing Strain Background

1. *HMG1* (3-hydroxy-3-methyl-glutaryl CoA reductase) gene from *S. cerevisiae* is 3165 bp in length. It contains N-terminal regulatory region (corresponds to 1-1590 bp) and C-terminal catalytic region (corresponds to 1591-3165 bp). The 525 aa (1575 bp (from 1591 to 3165 bp)) C-terminal catalytical region (*tHMG1*) is PCR amplified from *S. cerevisiae* genomic DNA using primers *tHMG1*-FP and *tHMG1*-RP primers (Table 1). *tHMG1* PCR product is restriction digested with *Bam*HI and *Xma*I and purified by gel extraction kit. Perform ligation of restriction enzyme digested and column purified pRS313TEF vector (yeast centromeric vector) and PCR product and kept at 16 °C for 16 h.
2. Transform the ligated reaction mixture in competent *E. coli* DH5α cells and transformants were screened for positive clones by restriction enzyme digestion. Glycerol stock was made in 50% glycerol and kept at −80 °C. Isolate the plasmid from *E. coli* transformants containing pRS313TEF-tHMG1 and quantify by NanoDrop. Transform the plasmid-pRS313TEF or pRS313TEF-tHMG1 in yeast transformants obtained in previous steps using protocol mentioned above (Subheading 3.1.2). Select yeast transformants on SD plates supplemented with appropriate amino acids but lacking uracil, tryptophan and histidine.

3.2.2 Validation of Screen from Visual Color of Dilution Spotted Transformants Containing tHMG1

1. Inoculate the yeast transformants obtained in previous step (Subheading 3.2.1) in 5 mL SD liquid media without uracil, tryptophan and histidine and incubate at 30 °C with orbital agitation 200 rpm for overnight.
2. After the overnight growth measure the optical density of culture at 600 nm (OD_{600}). Reinoculate the culture into 10 mL fresh SD liquid media supplemented with appropriate amino acids but lacking uracil, tryptophan, and histidine to

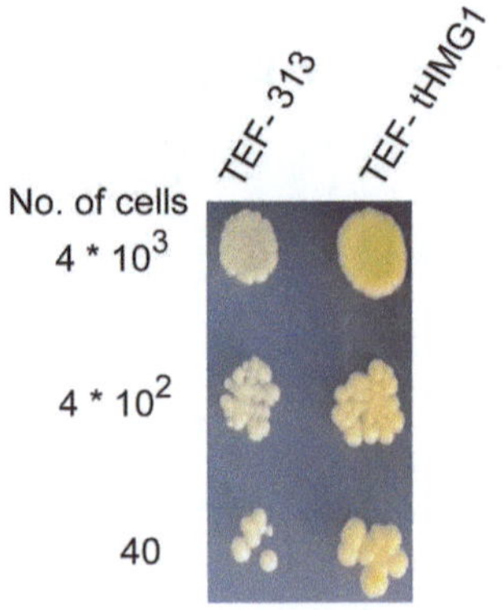

Fig. 3 Illustration of a result obtained in validation of developed genetic screen. Developed genetic assay strain is transformed with either pRS313TEF or pRS313TEF-tHMG1 and transformants were dilution spotted (at OD_{600} 0.02, 0.002 and 0.0002) on SD agar plates supplemented with appropriate amino acids but lacking uracil, tryptophan and histidine

have final $OD_{600} = 0.1$ and incubate at 30 °C till OD_{600} reaches to 0.6–0.8.

3. Harvest the yeast cells by centrifugation at 2800 × *g* for 3 min and discard the supernatant and wash the pellet with autoclaved distilled water.
4. Resuspend the pellet in 1 mL of autoclaved distilled water and measure the optical density OD_{600}.
5. Dilute the culture to make different dilutions having OD_{600}–0.02, 0.002, 0.0002.
6. Then spot 10 μL of different dilutions on SD agar plate supplemented with appropriate amino acids but lacking uracil, tryptophan and histidine.
7. Incubate plates at 30 °C for 2–3 days. Increase in color and growth is observed.

Illustrative results are provided in Fig. 3.

3.3 Mutagenesis and Screening of Mutants that Lead to Increased Isoprenoid Flux

SPT15 is a global TATA binding protein present in *S. cerevisiae* and encodes a protein of 240 aa.

3.3.1 Cloning of Chosen Target Gene- SPT15

1. To clone the coding region of *SPT15* gene (723 bp), perform PCR amplification using *S. cerevisiae* genomic DNA as a template using primers *SPT15*FP and *SPT15*RP (Table 1).
2. Purify the PCR amplified product using the gel extraction kit and perform the restriction digestion with *Bam*HI and *Xho*I of purified PCR product and pRS313TEF vector at 37 °C for 3 h.
3. Run the restriction enzyme digested PCR product and pRS313TEF vector on 1.2% agarose gel and purify using the

gel extraction kit. Perform ligation of vector and PCR product and kept at 16 °C for 16 h.

4. Transform the ligation mixture in competent *E. coli* DH5α cells and spread it on LB amp plates. Keep the plates at 37 °C.
5. Screen positive clones by restriction enzyme digestion with *Bam*HI and *Xho*I. Make glycerol stock of positive clones containing pRS313TEF- SPT15 in 50% glycerol and kept it at −80 °C.
6. Purify the plasmid pRS313TEF-SPT15 using kit and check the concentration of purified plasmid by NanoDrop.

3.3.2 In Vitro Mutagenesis of Plasmid DNA by Hydroxylamine

1. To the 10 μg of plasmid DNA (pRS313TEF-SPT15) in 1.5 mL microcentrifuge tube, add 500 μL freshly prepared hydroxylamine solution (as described in Subheading 2.3.4).
2. Seal the microcentrifuge tube with Parafilm and keep the solution at 37 °C for 20 h.
3. Purify hydroxylamine mutagenized plasmid DNA using gel extraction kit and elute the plasmid DNA in 100 μL of autoclaved distilled water and check the concentration of purified mutagenized plasmid library by NanoDrop.

3.3.3 Transformation and Screening of the Mutagenized Library

1. A single colony obtained from yeast transformation in previous steps (Subheading 3.1.2) is inoculated in 5 mL of SD liquid media supplemented with appropriate amino acids but lacking uracil and tryptophan and incubate at 30 °C with orbital shaking 200 rpm for overnight.
2. After the overnight growth, measure the optical density of culture at 600 nm (OD_{600}). Reinoculate the culture into 100 mL fresh SD liquid media supplemented with appropriate amino acids but lacking uracil and tryptophan to have final $OD_{600} = 0.1$ and incubate at 30 °C with orbital agitation 200 rpm.
3. Harvest the culture when OD_{600} of the culture reaches to approx. 1.0 and transform the hydroxylamine mutagenized plasmid library as described above (in Subheading 3.1.2). Transform approximately 600–700 ng of plasmid DNA per yeast transformation. Select approximately 10,000 yeast transformants on SD plates supplemented with appropriate amino acids but lacking uracil, tryptophan and histidine.
4. Screen the yeast colonies on the basis of increased color as compared to control strain containing wild type pRS313TEF-SPT15. Patch the colonies showing increased color on SD plates supplemented with appropriate amino acids but lacking uracil, tryptophan and histidine and incubate the plates at 30 °C.

5. Purify the plasmids from the yeast colonies patched in previous step using the following protocol.

Yeast Plasmid Isolation and Confirmation by Retransformation into Yeast

1. Inoculate the patched colonies in 10 mL SD liquid media supplemented with appropriate amino acids, but lacking uracil, tryptophan, and histidine and incubate at 30 °C with orbital shaking 200 rpm for 12–16 h.
2. Centrifuge the culture at 2800 × *g* for 3 min.
3. Discard the supernatant and wash the pellet with autoclaved distilled water.
4. Perform the plasmid isolation using plasmid miniprep kit. Briefly resuspend the pellet in 250 μL in resuspension solution and then add 250 μL Lysis solution and 250 μL glass beads.
5. Vortex it on bead beater for 30 s and incubate it on ice for 30 s (*see* **Note 6**).
6. To the microcentrifuge tube, add 350 μL chilled Neutralization buffer. Mix the solution by inversion and incubate it on ice for 5 min.
7. Centrifuge the tubes at 11,000 × *g* for 10 min.
8. Transfer the supernatant to column and centrifuge at 11,000 × *g* for 90 s.
9. Discard the flow through and place the column back to collection tube.
10. Wash the column with 500 μL wash buffer. Centrifuge the column at 11,000 × *g* for 90 s.
11. Discard the supernatant and place the column back to the collection tube and centrifuge at 11,000 × *g* for 2 min.
12. Place the column in fresh microcentrifuge tube and elute in 30 μL of elution buffer.
13. Transform the eluted plasmid in competent *E. coli DH5α* cells and select the transformants on LB Amp plates.
14. Perform plasmid isolation from clones and screen for pRS313TEF-SPT15 plasmid by restriction digestion.
15. Transform the positive purified mutagenized plasmid pRS313TEF-SPT15 obtained in previous step in yeast transformants obtained previously in Subheading 3.1.2.
16. If the transformants show increased color, then perform the sequencing of the plasmid—pRS313TEF-SPT15.
17. Analyze the sequence to find out mutation in the gene and perform subcloning of gene in fresh vector background.

3.4 To Evaluate If the Isolated Mutants Are Increasing the Flux Specifically in the Carotenoid Pathway or Are Increasing General Flux in Isoprenoid Pathway

The cDNA clone for α-farnesene synthase (*At*FS) (Locus AT4G16740, Clone no. U88221) with size 1000 bp is obtained from TAIR database, USA.

3.4.1 Cloning of α-Farnesene Synthase

1. Perform PCR amplification using *At*FS cDNA clone as a template and *At*FS FP and *At*FS RP (containing *Bam*HI and *Xho*I restriction site respectively) as primers mentioned in Table 1.
2. Perform the restriction enzyme digestion of amplified PCR product and pRS315TEF vector (yeast single copy centromeric vector) with *Bam*HI and *Xho*I at 37 °C and purify the digested PCR product and vector using gel extraction kit.
3. Perform ligation reaction of vector and PCR product purified in previous step using T4 DNA Ligase at 16 °C for 16 h. Transform the ligation mixture in competent *E. coli* DH5α cells and select the transformants on LB Amp plates and screen the positive clones by restriction digestion.

3.4.2 Construction of α-Farnesene Producing *S. cerevisiae* Strain

1. Transform the positive clone of pRS315TEF- *At*FS in ABC 276 strain using the protocol as described above (in Subheading 3.2.1) and select the transformants on the SD agar plates supplemented with appropriate amino acids but lacking leucine.
2. In these transformants, transform the pRS313TEF/ pRS313TEF-tHMG1 /pRS313TEF- SPT15 mutant plasmids isolated previously (in Subheading 3.3).
3. Select the transformants on SD agar plates supplemented with appropriate amino acids but lacking leucine and histidine.

3.4.3 Quantification of α-Farnesene

1. A single colony obtained in previous step (Subheading 3.4.2) is inoculated in 5 mL of SD media supplemented with appropriate amino acids but lacking leucine and histidine and incubate at 30 °C with orbital shaking 200 rpm for 16 h.
2. After the overnight growth, measure the optical density of culture at 600 nm (OD_{600}). Reinoculate the culture into 100 mL fresh SD liquid media supplemented with appropriate amino acids but lacking leucine and histidine to have final $OD_{600} = 0.05$ and incubate at 30 °C with orbital agitation 200 rpm.
3. Overlay the culture with 10% (v/v) dodecane layer when OD_{600} of the culture becomes 0.6–0.8.
4. After 48 h of the growth, take the dodecane phase of two-phase culture and centrifuge at 2800 × *g* for 5 min at room temperature.
5. Take 1 μL of the dodecane phase and perform GC-FID analysis using the following conditions (*see* **Note 7**):

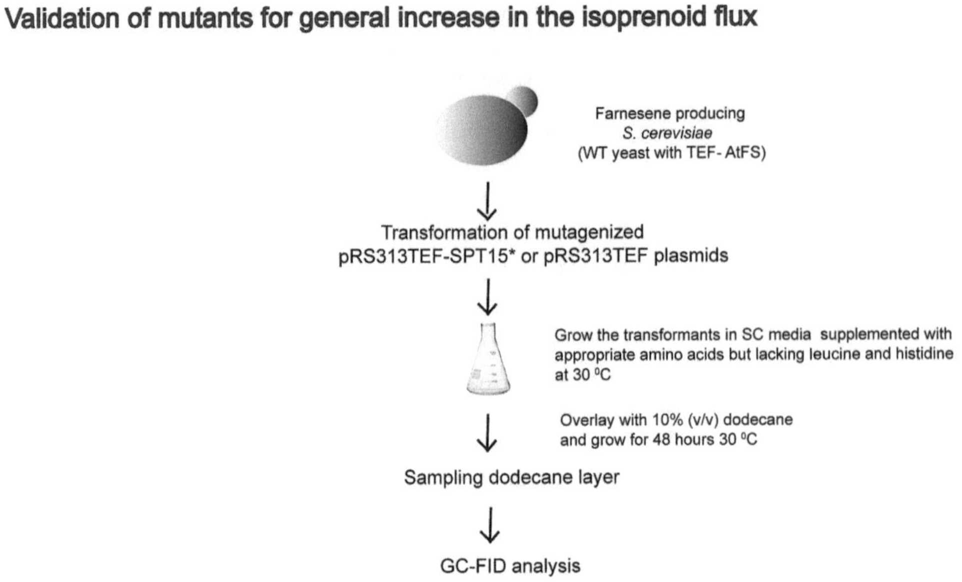

Fig. 4 Validation of mutants for increasing the general flux in isoprenoid pathway in *Saccharomyces cerevisiae*

Split ratio—1:10.

Oven temperature—80 °C for 1 min and increased to 250 °C at a rate of 10 °C/min and hold at 250 °C for 1 min.

Temperature of detector—260 °C.

6. Quantify the amount of α-farnesene produced by yeast strain by calculating it from standard curve prepared from standard trans β-farnesene (Cat. no. 73492, Sigma Aldrich, India).
7. Concentration of α-farnesene (μg/mL) = Peak Area/ 3000810.

Schematic representation of the procedure for validation of mutants for increasing the general isoprenoid flux is given in Fig. 4.

4 Notes

1. Autoclave all media and solutions at specified conditions, otherwise media will get contaminated.
2. Prepare LB Amp agar plates when the LB Agar media cools down to approximately 50–60 °C, otherwise ampicillin will get degraded.
3. Prepare 50% Poly Ethylene Glycol solution in 1× TE, Take half the volume of 1× TE to be prepared and mix PEG slowly while

heating on magnetic stirrer. PEG will not dissolve without heating the solution. Volume of the solution will rise after mixing of PEG, therefore make up the volume of solution after dissolving PEG.

4. Always prepare the fresh hydroxylamine solution and keep it on ice.
5. Perform the ligation reaction at 16 °C for 16 h.
6. Perform lysis of yeast cells on bead beater by performing bursts for 30 s and keeping on ice for 30 s, in yeast plasmid isolation protocol.
7. Carefully collect the dodecane solution layer from two phase culture and centrifuge the samples to remove the cells that come along with dodecane layer. Inject only dodecane layer in GC-FID.

Acknowledgments

MW acknowledges support from university grant commission for senior research fellowship. AKB is a JC Bose fellow of the Department of Science and Technology (DST), Govt. of India. The project was funded by grant (BT/PR13900/BBE/117/34/2015) from Department of biotechnology (DBT), Govt. of India. We would like to thank Dr. R. Vijayanand and Dr. Asim Chaudhury from Department of chemistry, IISER Mohali for their help.

References

1. Bohlmann J, Keeling CI (2008) Terpenoid biomaterials. Plant J 54(4):656–669. https://doi.org/10.1111/j.1365-313X.2008.03449.x
2. Emmerstorfer-Augustin A, Moser S, Pichler H (2016) Screening for improved isoprenoid biosynthesis in microorganisms. J Biotechnol 235:112–120. https://doi.org/10.1016/j.jbiotec.2016.03.051 S0168-1656(16)30167-5 [pii]
3. Ignea C, Trikka FA, Nikolaidis AK, Georgantea P, Ioannou E, Loupassaki S, Kefalas P, Kanellis AK, Roussis V, Makris AM, Kampranis SC (2015) Efficient diterpene production in yeast by engineering Erg20p into a geranylgeranyl diphosphate synthase. Metab Eng 27:65–75. https://doi.org/10.1016/j.ymben.2014.10.008 S1096-7176(14)00138-4 [pii]
4. Lee PC, Petri R, Mijts BN, Watts KT, Schmidt-Dannert C (2005) Directed evolution of Escherichia coli farnesyl diphosphate synthase (IspA) reveals novel structural determinants of chain length specificity. Metab Eng 7 (1):18–26. https://doi.org/10.1016/j.ymben.2004.05.003 S1096-7176(04)00080-1 [pii]
5. Leonard E, Ajikumar PK, Thayer K, Xiao WH, Mo JD, Tidor B, Stephanopoulos G, Prather KL (2010) Combining metabolic and protein engineering of a terpenoid biosynthetic pathway for overproduction and selectivity control. Proc Natl Acad Sci U S A 107 (31):13654–13659. https://doi.org/10.1073/pnas.1006138107 1006138107 [pii]
6. Ohnuma S, Nakazawa T, Hemmi H, Hallberg AM, Koyama T, Ogura K, Nishino T (1996) Conversion from farnesyl diphosphate synthase to geranylgeranyl diphosphate synthase by random chemical mutagenesis. J Biol Chem 271 (17):10087–10095
7. Ohnuma S, Suzuki M, Nishino T (1994) Archaebacterial ether-linked lipid biosynthetic gene. Expression cloning, sequencing, and

characterization of geranylgeranyl-diphosphate synthase. J Biol Chem 269(20):14792–14797

8. Schmidt-Dannert C, Umeno D, Arnold FH (2000) Molecular breeding of carotenoid biosynthetic pathways. Nat Biotechnol 18 (7):750–753. https://doi.org/10.1038/77319
9. Umeno D, Tobias AV, Arnold FH (2002) Evolution of the C30 carotenoid synthase CrtM for function in a C40 pathway. J Bacteriol 184 (23):6690–6699
10. Ozaydin B, Burd H, Lee TS, Keasling JD (2013) Carotenoid-based phenotypic screen of the yeast deletion collection reveals new genes with roles in isoprenoid production. Metab Eng 15:174–183. https://doi.org/10.1016/j.ymben.2012.07.010 S1096-7176(12)00081-X [pii]
11. Yuan J, Ching CB (2014) Combinatorial engineering of mevalonate pathway for improved amorpha-4,11-diene production in budding yeast. Biotechnol Bioeng 111(3):608–617. https://doi.org/10.1002/bit.25123
12. Verwaal R, Wang J, Meijnen JP, Visser H, Sandmann G, van den Berg JA, van Ooyen AJ (2007) High-level production of beta-carotene in *Saccharomyces cerevisiae* by successive transformation with carotenogenic genes from *Xanthophyllomyces dendrorhous*. Appl Environ Microbiol 73(13):4342–4350. https://doi.org/10.1128/AEM.02759-06 AEM.02759-06 [pii]
13. Wadhwa M, Bachhawat AK (2016) A genetic screen for increasing metabolic flux in the isoprenoid pathway of *Saccharomyces cerevisiae*: Isolation of SPT15 mutants using the screen. Metab Eng Commun. https://doi.org/10.1016/j.meteno.2016.05.004
14. Wadhwa M, Bachhawat AK (2018) Increased production of isoprenoid in *Saccharomyces cerevisiae* by carotenoid optimization and screening. US Patent number (15/066296)

INDEX

A

B

C

Christine Nicole S. Santos and Parayil Kumaran Ajikumar (eds.), *Microbial Metabolic Engineering: Methods and Protocols*, Methods in Molecular Biology, vol. 1927, https://doi.org/10.1007/978-1-4939-9142-6,

D

E

F

O

P

Q

R

S

MIX
Papier aus verantwortungsvollen Quellen
Paper from responsible sources
FSC® C105338

If you have any concerns about our products,
you can contact us on
ProductSafety@springernature.com

In case Publisher is established outside the EU,
the EU authorized representative is:
Springer Nature Customer Service Center GmbH
Europaplatz 3, 69115 Heidelberg, Germany

Printed by Libri Plureos GmbH
in Hamburg, Germany